HUMAN GENETICS

Readings on the Implications of Genetic Engineering

THOMAS R. MERTENS

JOHN WILEY & SONS, INC.

New York/London/Sydney/Toronto

Library of Congress Cataloging in Publication Data:

Mertens, Thomas Robert, 1930—   comp.
  Human genetics.

  Bibliography: p.
    1. Genetic engineering—Social aspects. 2. Human genetics—Social aspects. I. Title. [DNLM: 1. Genetic intervention—Collected works. 2. Genetics, Human—Collected works. 3. Social problems—Collected works. QH442 M575h]
QH442.M47   301.24'3   74-30471
ISBN 0-471-59628-0

Printed in the United States of America
10 9 8 7 6 5 4 3 2 1

Preface

Since Correns, De Vries, and Von Tschermak rediscovered Mendel's basic genetic research in 1900, genetics has become a significant, unifying discipline in biology. It provides the foundation for explaining the mechanism of evolution and the transmission of traits at the cellular and organismic levels. During the last quarter century biochemical and microbial geneticists have made major and often dramatic contributions to the advancement of molecular biology. As you will learn here, some of these advances made possible new developments in human genetics.

Until recently, advances in human genetics were not as rapid or dramatic as the progress in other fields of genetic research. Progress was hampered by the very nature of the human species: the long generation time, the comparatively small size of human families, and the impossibility of conducting experimental matings. However, in the 75 years since the rediscovery of Mendel's work, human genetics has developed into a fascinating science. Because recent progress has been rapid, many scientists predict the use of genetic knowledge to manipulate human reproduction and heredity. These manipulations are popularly called "genetic engineering" and have as their goal the correction of genetic defects and the ultimate alteration of the human gene pool in order to direct the course of human evolution.

Major social, legal, and ethical problems have risen from recent advances in human genetics and from the technology that these advances will make possible. Some of the current problems concern developments in genetic counseling, sperm banks, amniocentesis, artificial insemination by donors (AID), and the abortion of genetically defective foetuses. The problems facing us in the near future could be much more complicated. They could include the cloning of human eggs with the consequent asexual reproduction of numerous genetically indentical individuals, genetic surgery or gene repair (perhaps through viruses), legislative regulation of human reproduction, and ultimately human control over the genetic future of *Homo sapiens.*

Some authorities argue that most of these problems will arise in the future, that they may never materialize, and that we can deal with them if they do arise. Others suggest that the problems are of the same magnitude as the ones that accompanied the development of atomic warfare. Society was not prepared for the social and ethical problems posed by the atomic age. Should we also enter a genetic age unprepared? Many biologists, educators, sociologists, and theologians suggest that we should deliberate about future contingencies while we have the time, instead of being forced to make decisions that have profound biological, social, and ethical implications under the pressure of some "genetic bomb."

I think that some forethought should be given to these problems. Certainly, college biology majors, and especially future secondary school biology teachers, need to be informed about current developments in human genetics and the implications of these developments for genetic engineering. If intelligent decisions on these problems are to be made, a scientifically literate citizenry is an absolute necessity.

Most of society has the potential for developing scientific literacy. The growing popular interest in human genetics and genetic engineering is reflected in these current developments: (1) the increasing number of articles on the subjects in popular magazines and newspapers, (2) the willingness of informed scientists, physicians, sociologists, and theologians to discuss the issues so that the public can understand them, and (3) the proliferation of college and university courses that study human genetics at a level that the nonbiologist can grasp.

The first two developments have made possible this book of readings on genetic engineering. The selectious are from a variety of sources: scientific and medical journals, journals for science educators, and from such popular magazines as *Saturday Review.* I chose articles because of their biological significance, their readability, and their meaningfulness for the informed layman. They are provocative and often controversial. Ultimately they will help the reader to better understand and cope with some of the problems with which he is likely to be confronted during the last quarter of this century.

MUNCIE, INDIANA, 1974   THOMAS R. MERTENS

Contents

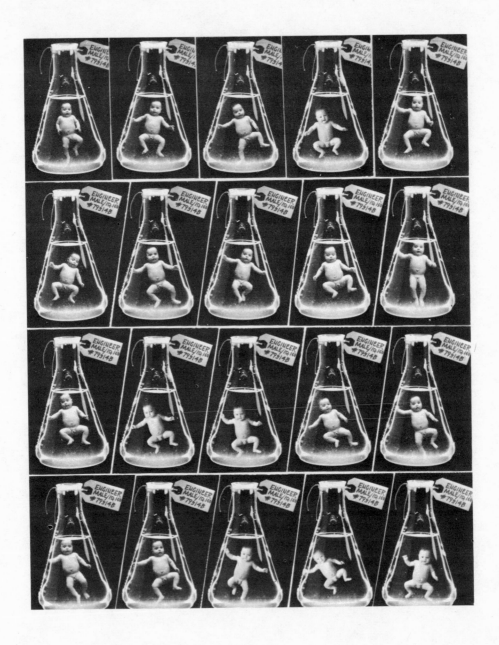

I
GENETIC AND REPRODUCTIVE ENGINEERING

Brave New World or
Grave New World?

In late June, 1974 *Time* magazine published a series of beautiful photographs of ovulation, fertilization, and early embryonic development that were taken *in vivo* by Dr. Motoyuki Hayashi of Tokyo's Toho University School of Medicine. Although rabbits and monkeys were used for the photographs of ovulation and fertilization, sequences showing fetal development were photographed in the uterus of a human female. Furthermore, there is no question that the technology used for making the photographs in animals could also be applied successfully to human females.

In mid-July, 1974 British medical authorities announced the birth of what the popular press called "the first test tube babies." Although this dramatic description may imply more than the actuality, the development signaled by the announcement is still significant: Eggs of human females were removed surgically from the ovaries, fertilized *in vitro* (presumably by sperms from each woman's husband) and, after approximately a week's development *in vitro* the developing embryos were implanted in the receptive uteri of the three women in question. Subsequent development proceeded normally *in utero,* and in due course three apparently normal children were born.

Although these two recent developments are not concerned directly with human genetics, they are the kinds of advances that fall under the inclusive title, "genetic and reproductive engineering." They are also the kinds of developments that capture the imagination of the general public and raise social, ethical, and legal questions that concern scientists, physicians, sociologists, lawyers, and theologians. To intervene in the human reproductive process, to alter the genetic constitution of an individual, and to modify the genetic makeup of the human species are awe-inspiring prospects that are coming closer to reality. Although some of the technology of genetic and reproductive engineering has yet to be perfected, society would do well to consider the social and ethical questions raised by genetic engineering so that it does not have to make important decisions under pressure.

In the introductory article in this book, Caryl Rivers, an instructor at the Boston University School of Public Communications and a free-lance writer, discusses some of the discoveries that suggest that sophisticated genetic engineering techniques may soon be possible. She notes that even such simple procedures as genetic screening and counseling raise ethical questions, while such proposed procedures as cloning and gene surgery pose enormously complex and frustrating moral issues. If there are problems created by an attempt to improve the genetic constitution of one individual, to correct a genetic error he possesses, or to prevent his transmitting the error to his offspring—how much more complicated are the questions created by the possibility of regulated human reproduction with the goal of altering the human gene pool and thus the evolutionary future of the entire species!

Hopefully Caryl Rivers' article, by suggesting some of the techniques that might be used to accomplish genetic engineering and some of the problems that would result if these techniques were applied, will whet your intellectual appetite and cause you to probe further in this book. The fact that Rivers' article was published in a popular magazine such as *Saturday Review* suggests that the problems of genetic engineering are of concern to a broad sector of the educated public in the United States. Certainly in the next quarter century we can expect to see advances in genetic and reproductive technology force us into a position requiring many difficult decisions. If scientists, physicians, lawyers, and theologians continue to share their insights into these problems with all of society, perhaps wise decisions can be made. If you wish to participate intelligently in making these decisions, you will need to keep abreast of many of the developments described by Caryl Rivers and expanded upon in other articles in this book.

1
Genetic Engineering Portends A Grave New World

CARYL RIVERS

A society matron proudly introduces her two sons and her daughter. They are her children by every biological rule but one: She has never been pregnant. Sex cells from her body and her husband's body were implanted three times in the womb of a "proxy mother," who was paid a union wage to carry each fetus and give birth.

An astronaut is carried aboard a space vehicle destined to probe the outer limits of this galaxy. He has no legs. Legs would be only an inconvenience during the years in which the astronaut will be confined to his spaceship. For this reason he was programed to be born legless.

A Latin American dictator has some skin tissue scraped from his left arm. Nine months later, five hundred babies emerge from a factory that contains five hundred artificial wombs. The babies are genetic carbon copies of the dictator.

Although fiction today, these events could become realities with relatively slight advances in man's most adventurous and morally complex science, genetic engineering.

This new science is experimenting with a technique that would make possible the manipulation of an embryo during gestation so as to change its physical characteristics. It may offer an alternative to natural human reproduction—a process that would allow the implantation of a fertilized ovum in the womb of a "host" mother or in an artificial womb inside a laboratory. Not too far off, according to specialists in the field, is the possibility of creating children with only one parent who will be biological duplicates of that single parent. Genetic engineering, in short, is on the brink of revolutionizing the traditional concepts of man, God, and creation.

The moral questions posed by recent advances in this and related life sciences are no longer speculative. They have to be faced as practical realities, and, in parts of the scientific community, this is now being done. Indeed, a heated debate is in progress. Some scientists encourage genetic experimentation, putting their faith in man's rational power. They say, in effect, that the only thing too sacred to tamper with is scientific investigation itself. Others, however, point to numerous instances in which technology has outrun man's wisdom (nuclear stockpiles, for example), and they warn that if guidelines in genetic technology are not immediately set forth, geneticists may bring into being a terrifying, uncontrollable Huxleian nightmare.

Those who urge restraint say it is past time to start thinking about controls for genetic technology. Scientific advances have already made it possible to change fundamentally the human reproductive process. The Caesarean section is now a common operation and constitutes a basic tampering with the way in which babies are born. There have already been successful blood transfusions to unborn children in cases of Rh blood incompatibility. An estimated 25,000 women whose husbands are sterile resort to artificial insemination in this country each year, and more than a third of them give birth as a result.

Test-tube fertilization of a human ovum is already a reality. In 1950 Dr. Landrum Shettles of Columbia University said he had achieved in vitro —outside the womb—fertilization of an ovum and had maintained the embryo's life for six days. In 1961 Dr. Daniele Petrucci of the University of Bologna claimed he had fertilized an egg in vitro and that the embryo lived for twenty-nine days, although it became enlarged and deformed. His work was condemned by Pope John XXIII.

Perhaps the most advanced work in the field is being done at Cambridge University in England by Drs. Robert G. Edwards and P. S. Steptoe. They have produced the best evidence to date of true in vitro fertilization and are presently attempting to implant the fertilized egg in the uterus of the donor. They obtain the ova for their experiments from volunteer women who are seeking to overcome infertility. One woman, for example, had blocked Fallopian tubes that made normal conception impossible.

Once an ovum can be successfully implanted in the womb, the way is open for "proxy mothers." A fertilized ovum could be removed from the womb of a woman after she had conceived and then be implanted in the uterus of another woman who would ultimately give birth. The child, of course, would carry the genetic identity of its true parents. The proxy mother would be only a temporary host, with no genetic relationship to the child. Technology could go one step further and simply eliminate the process of pregnancy and childbirth altogether. A human embryo could be removed from the uterus and placed in an artificial womb. A prototype of such a device was constructed by Dr. Robert Goodlin at Stanford University. It is a pressurized steel chamber containing an oxygen-rich saline solution that would push the oxygen through the body of the fetus. A fetus put into the chamber, however, would perish from its own wastes, which the placenta draws off in natural pregnancy. Dr. Goodlin is doing no further work in the area and his announced that he does not intend to resume his experiments with the artificial womb.

At the National Heart Institute at Bethesda, Maryland, Drs. Warren Zapol and Theodore Kolobow are placing lamb fetuses in a liquid solution and attaching the umbilical cords to machinery that contains an artificial lung, a pump, and a supply of nutrients. The fetuses survive several days before "dying" of cardiac arrest.

Such *ex utero* gestation might offer some obvious advantages. If made common practice, doctors would be able to monitor the growth of the embryo closely. A new specialist, the "fetologist," would check to ensure normal development and catch possible medical problems in the earliest stages. The trauma of birth, from which some babies die or suffer injury, would be avoided altogether. *Ex utero* development would make more likely manipulation of the developing fetus to change its sex, size, and other physical characteristics.

French biologist Jean Rostand speculated that, once the size of the human brain is not limited by the size of the female pelvis, it might be possible to double the number of fetal brain cells. Would this produce superintelligent people or monstrous misfits? Where will the line be drawn between legitimate medical practices and a Frankenstein-like toying with the human condition? And who will draw the line?

Another possibility of genetic engineering that raises serious moral questions is the creation of what are called clones—carbon copies of a human being already in existence, resulting from a process much more revolutionary than the merging of sperm and egg in the test tube.

Normally, the sexual process by which life begins guarantees the diversity of the species. The child inherits characteristics from both the mother and the father so that, while he may resemble either or both of them, he is unique. There are two kinds of cells in the human body—sex cells (ova and sperms) and body cells. Each body cell has a nucleus containing forty-six characteristic-determining chromosomes. Sex cells have only half that number. A merger of two sex cells is required for the resulting zygote to have a complete set of chromosomes and to begin dividing.

But human reproduction could take another route. Body cells contain the requisite number of chromosomes to start a new individual, but they cannot divide the way the fertilized egg does. The body cells have very specialized jobs to do. Some go into making hair, others teeth, others skin, and so forth. Only the part of the body-cell mechanism that is useful in its specific job gets "switched on." If a body cell could receive a signal to switch on all of its mechanisms, it would divide and subdivide and multiply, finally developing into a new human being.

Accordingly, if such a "switching on" technique were developed, a body cell could be taken from a donor—scraped from his arm, perhaps—and be chemically induced to start dividing. The cell could be implanted in an artificial womb or in the uterus of a woman, where, presumably, it would develop like a normal fetus. The baby would be genetically identical to the donor of the cell—his twin, a generation removed. It would have only one true parent. Its "mother" would be, like the proxy mother, only a temporary host.

The possibilities of human cloning are enough to startle even a science-fiction writer. Societies would be tempted to clone their best scientists, soldiers, and statesmen. An Einstein might become immortal through his carbon copies. Or a Hitler. What better way for a dictator to extend his power beyond the grave? A carbon copy might be the ultimate expression of human egoism.

Clones have already been produced from carrots and frogs. Dr. J. B. Gurdon of Oxford University produced a clone from an African clawed frog by taking an unfertilized egg cell from a frog and destroying its nucleus by radiation. He then replaced it with the nucleus of a cell from the intestines of a tadpole. The egg began to divide as if it had been normally fertilized. The result: a twin of the donor tadpole.

Some scientists estimate that human cloning will be possible by the end of the 1970s. If human cloning were ever practiced on a wide scale, it would drastically affect the course of human evolution. Nobel Prize-winning geneticist Dr. Joshua Lederberg has said he can imagine a time in the future when "clonishness" might replace the present dominant patterns of nationalism and racism.

Cloning is not the only avenue man might take to redesign himself. Advances in molecular biology could lead to genetic surgery: the addition of genes to or removal of them from human cells. In experiments with mouse cells, scientists added healthy new genes to a defective cell. The new genes not only corrected an enzyme deficiency in the cell but were duplicated when healthy new cells began to divide. It is now possible, in principle, to remove defective cells from a human patient, to introduce new genes, and to place the "cured" cells back into the patient.

Genetic surgery might also be put to more ambitious uses. J. B. Haldane, the late, renowned British geneticist, was one of the first to speculate and predict that men would tailor-make new kinds of people for space travel. He suggested the legless man referred to earlier, who would be ideally suited to living in the cramped quarters of a space capsule for long journeys; men with prehensile feet and tails for life on asteroids where low gravity makes balance difficult; and muscular dwarfs to function in the strong gravitational pull of Jupiter.

Genetic engineering might also bring to life one of the mixtures of man and animal known in all the myths of man: the Minotaur, the Centaur, and the Gorgons. So far, however, human and animal chromosomes have been successfully mingled only in tissue-culture experiments.

Modern medicine may be able to produce not only man-animal combinations but a man-machine combination for which a name has already been coined: cyborg. We are familiar with the use of artificial material in the human body: plastic arteries, synthetics to replace damaged bone, or even an artificial heart. Who will draw the dividing line between man and robot?

There are few existing guidelines for the manipulation of "living" material and human beings. Reports from Britain describe experiments in which pregnant women awaiting abortion have been exposed to certain kinds of sound waves to determine if there will be extensive chromosome damage in the fetus. Is it immoral to damage a fetus for experimental purposes, even though it will be aborted? Does an embryo "grown" in a laboratory have any legal rights?

Even the practice of genetic counseling—guidance based upon an individual's chromosomal make-up—which is becoming increasingly common, raises ethical questions. Would an individual who is marked as a carrier of deleterious genes carry a social stigma as well? Dr. Marc Lappé of the Institute of Society, Ethics, and the Life Sciences, a group set up to examine the moral questions posed by science, thinks he would.

"It could lead to a subtle shift in the way we identify people as human," he says. "We could say, 'Oh, he has an extra chromosome.' We could then identify him as being qualitatively different. He is somewhat imperfect, perhaps not as human as the rest of us."

Suppose, for example, a genetic counselor knows that a newborn baby has an extra Y chromosome. There is evidence that the extra Y chromosome predisposes in individual to aggressive behavior. Should that become part of his medical record? What impact might that knowledge have on a school principal if a child with an extra Y becomes involved in typical childish pranks? Should the parents know, or would the knowledge have an adverse effect on the way they bring up the child?

The agonizing moral question posed by almost all aspects of genetic engineering has divided the scientific community. Hearing the arguments among scientists, one is reminded of the story about two men in a jail cell. One looked out and saw the mud; the other saw the stars.

Dr. Robert Sinsheimer of the California Institute of Technology sees the promise: "For the first time in all time a living creature understands its origin and can undertake to design its future. Even in the ancient myths man was constrained by his essence. He could not rise above his nature to chart his destiny."

Dr. Sinsheimer goes on to say that those who oppose genetic engineering "aren't among the losers in the chromosomal lottery that so firmly channels our human destiny." Repugnance to advances in genetic technology "isn't the response of four million Americans born with diabetes, or the two hundred fifty thousand children born in the United States every year with genetic diseases, or the fifty million Americans whose I.Q.s are below ninety."

Dr. Salvador Luria of MIT, a Nobel laureate, takes the opposite viewpoint. "We must not ignore the possibility that genetic means of controlling human heredity will become a massive means of human degradation. Huxley's nightmarish society might be achieved by genetic surgery rather than by conditioning and in a more terrifying way, since the process would be hereditary and irreversible."

The work of Steptoe and Edwards in Cambridge has been extremely controversial among scientists. Is implantation of a fertilized ovum back into the uterus a desirable medical advance or the opening of a Pandora's box? Dr. Joseph Fletcher of the University of Virginia Medical School argued the former in a recent article that he published in the *New England Journal of Medicine.*

"It seems to me," he wrote, "that laboratory reproduction is radically human compared to conception by ordinary heterosexual intercourse. It is willed, chosen, purposed, and controlled, and surely these are among the traits that distinguish Homo sapiens from others in the animal [world], from the primates down. Coital reproduction is therefore less human than laboratory reproduction; more fun, to be sure, but with our separation of baby-making from love-making, both become more human, because they are matters of choice, not chance. I cannot see how humanity or morality [is] served by genetic roulette."

At the opposite pole, Dr. Leon Kass, the executive director of a committee on the life sciences and social policy for the National Academy of Sciences, argues that the further reproduction is pushed into the laboratory the more it becomes sheer manufacture.

"One can purchase quality control of the product only by the depersonalization of the process," he says, going on to ask, "Is there not some wisdom in the mystery of nature that joins the pleasure of sex, the communication of love, and the desire for children in the very activity by which we continue the chain of human existence? Is not human procreation, if properly understood and practiced, itself a humanizing experience?"

Dr. Kass points to the lonely depersonalized experiences old age and dying have become as a result of medical technology. The aged are kept alive but barely able to function, and they die in institutions surrounded by clacking machinery and uncaring strangers. He also believes that laboratory reproduction will deal a near fatal blow to the human family.

"The family is rapidly becoming the only institution in an increasingly impersonal world where each person is loved not for what he does, or makes, but simply because he is. Destruction of the family unit would throw us, even more than we are now, on the mercy of an impersonal, lonely present."

There are some members of the scientific community who take what might be called an "amber light" approach to genetic technology: Proceed with extreme caution.

Isaac Asimov, the biochemist who is also well known as a science-fiction writer, says we should be very sure what we're about before we start tinkering with genes. "We should intervene if we have reasonable suspicion that we can do so wisely. If our choice is doing nothing or doing something without knowing, we should do nothing. But wise change is better than no change at all. It's the same as changing the environment. Every time we build a dam there is a gain and there is a loss. What has usually happened is that we have considered the short-term gain without thinking what the long-term effect will be."

Dr. George Wald, a Nobel laureate at Harvard who has spoken out on a wide range of social issues, emphasizes the fact that every organism alive today represents an unbroken chain of life that stretches back some three billion years. That knowledge, he says, calls for some restraint. The danger he senses in genetic technology is a movement toward reducing man's unpredictability. "With animals we have abandoned natural selection for the technological process of artificial selection. We breed animals for what we want them to be: the pigs to be fat, the cows to give lots of milk, work horses to be heavy and strong, and all of them to be stupid. This is the process by which we have made all of our domestic animals. Applied to men, it could yield domesticated men."

Wald sees the trends of modern life working to erode one of the glories of being human: free will. "Free will is often inefficient, often inconvenient, and always undependable. That is the character of freedom. We value it in men. We disparage it in machines and domestic animals. Our technology has given us dependable machines and livestock. We shall now have to choose whether to turn it to giving us more reliable, efficient, and convenient men, at the cost of our freedom. We had better decide now, for we are already not as free as we once were, and we can lose piecemeal from within what we would be quick to defend in a frank attack from without."

Dr. Harold P. Green, a law professor at George Washington University, argues for the creation of an agency within the government whose sole task would be to make the case against technological advances. With any technology, he says, the benefits are immediate and obvious, and there are powerful vested interests to articulate them. The risks are more remote, and there are few people to argue these points. A prestigious new agency so mandated might right the present imbalance.

Senator Walter Mondale has proposed a fifteen-man presidential advisory committee that would spend two years looking into the moral problems posed by the life sciences. The commission would devise a structure to give society controls over such things as genetic technology. The proposal was approved by the Senate and, at last report, is languishing in a House committee.

Scientists themselves have shown growing awareness of the moral implications of their work. "Scientists have always had a supreme obligation to be concerned about the uses of their work," Isaac Asimov says. "In the past an ivory-tower scientist was just stupid. Today he's stupid and criminal."

But is it enough that scientists develop sensitivity to the moral complexities of the new technology? Can they alone grapple with choices so fundamental they could affect the future of human heredity? James Watson has called the idea that science must always move bravely forward "a form of laissez-faire nonsense dismally reminiscent of the creed that American business, if left to itself, would solve everybody's problems."

Seeing a child lying crippled by a genetic disease, one can't help thinking that, if such a scourge could be lifted from the children of the future, it would be worth the risk of any Brave New Worlds. It is probably unwise and perhaps impossible to barricade any street of scientific inquiry and say: No Admittance. But neither can we blind ourselves to the consequence of traveling that street until it is too late. Dr. Kass has summarized the question posed by genetic engineering: "Human heredity is intricate and mysterious. We must face the prospect of intervention with awe, humility, and caution. We may not know what the devil we are doing."

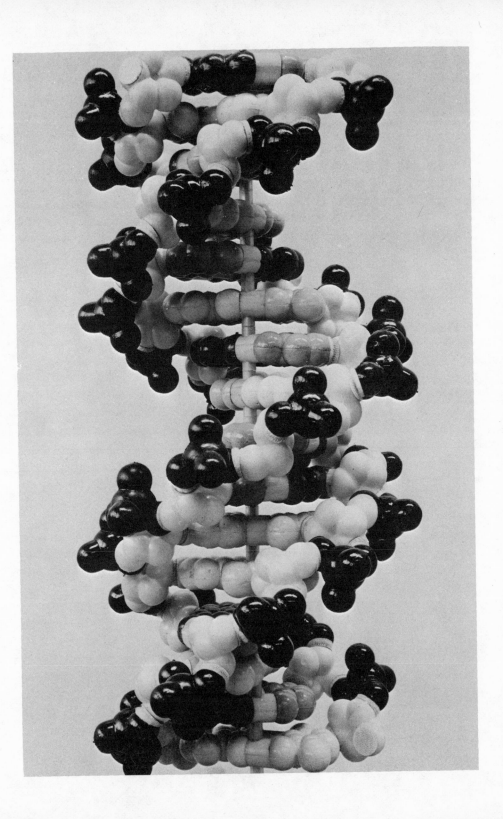

II PROGRESS IN HUMAN GENETICS

Foundations for Genetic Manipulation

Especially significant for appreciation of the remainder of the readings in this book is a review of the current status of human genetics, a discipline that has witnessed tremendous progress in the last two decades. In the first selection in this part of the book, medical geneticist Kurt Hirschhorn, M.D., brings us up to date with recent advances in human population genetics, cytogenetics, biochemical genetics, immunogenetics, somatic cell genetics and, finally, clinical or applied genetics. In his discussion of clinical genetics, Dr. Hirschhorn emphasizes the growing importance of genetic disorders to the daily practice of medicine. Not only will the practicing physician need to become familiar with these genetic disorders, but he will also need to seek and rely on advice and assistance from medical geneticists and genetic counselors in the diagnosis and treatment of these disorders.

Following Dr. Hirschhorn's article, we examine more intensively some of the dramatic discoveries in human cytogenetics. Especially striking advances in the study of human chromosomes began with the discovery by Tjio and Levan in 1956 that the human diploid chromosome number is 46 rather than 48 as had been generally accepted for over three decades. Dr. John H. Heller, president of the New England Institute presents a fascinating review of the most common human chromosome aberrations and their resulting physical and mental symptoms. Dr. Heller ends his article with a discussion of the 47,XYY chromosome complement and some of the ethical and legal implications raised by the putative behavioral characteristics of XYY men. This article, then, foreshadows much that follows in subsequent readings in the book relative to social, ethical, and legal implications of advances in human genetics and genetic engineering. Dr. Heller's article concludes with a comprehensive bibliography that will be useful to the reader who wishes to delve more deeply into the intricacies of human cytogenetics.

In the final selection in this part of the book Paul T. Libassi, a member of the staff of *Laboratory Management,* gives us some idea of the directions in which the science of human genetics is likely to be moving in the coming decade. Libassi describes the creation with federal government support of seven major research centers, each designed to specialize in the study of specific kinds of human genetic disorders. The need for these centers is documented and a brief account is given of the expectations that are held for them and for progress in human genetics.

2
Human Genetics

KURT HIRSCHORN, M.D.

The science of genetics is concerned with the inheritance of traits, whether normal or abnormal, and with the action of genes in the environment. This latter concept is of particular relevance to medical genetics, since the expression of genes can be modified by manipulating the environment. For example, were we to live in a society that takes in no dairy products, as well as avoids breast-feeding, we would not be aware of the inherited disease, galactosemia. Human genetics can be subdivided into population genetics, cytogenetics, biochemical genetics, immuno-genetics, somatic cell genetics and, of course, applied or clinical genetics. While there is a great deal of overlap among these subdivisions, I shall deal with each separately, both for the definition of terms and a review of their principles and applications.

Population Genetics

Population or mathematical genetics deals with the study of the mode of inheritance of traits and with the distribution of genes in populations. Since, with the exception of the sex chromosomes in the male, all chromosomes, the structures carrying the genes, exist in pairs, our cells

Reprinted by permission of the *Journal of The American Medical Association*. JAMA, Vol. 224 (5): 597-604, April 30, 1973.

contain two copies of each gene, which may be alike or may differ in their substructure and their product. Different forms of genes at the same locus or position on the chromosome are called alleles. If both copies of the gene are identical, the individual is homozygous, while if they differ, he is heterozygous. The exception to the rule that cells contain pairs of chromosomes applies to the gametes, sperm and ovum, which contain only single representatives of each pair of chromosomes and, therefore, of each pair of genes. When the two gametes join at fertilization, the new individual produced again has paired genes, one from the father and one from the mother. If a trait or disease manifests itself when the affected person carries only one copy of the mutant gene responsible, along with one normal allele, the mode of inheritance of the trait is called dominant. If two copies of the mutant gene are required for expression of the trait, the mode of inheritance is called recessive.

A family pedigree of a dominant trait will show parent-to-child transmission, that is, a parent and, on the average, half of his or her children, regardless of their sex, are affected. The normal children have received the affected parent's normal allele. Since many dominant traits are severely disabling (eg, Marfan's syndrome and osteogenesis imperfecta), fertility is markedly reduced, and many cases appear "sporadically without an affected parent" but due to a new mutation in one of the parents' germ cells. The frequency of such mutations (for many genes about 1×10^{-5} per generation) correlates somewhat with the father's age.

Recessive traits, on the other hand, are indicated by pedigrees in which two normal parents have an affected child; about one quarter of the siblings will be similarly affected. This is due to the fact that the two mutant genes in the affected offspring have come one each from the father and mother. Each parent is a heterozygote for the mutant gene and its normal allele and, therefore, does not express the trait. Random combination of the two types of allele from each parent will produce one quarter homozygous normal, one half heterozygous (clinically normal), and one quarter homozygous affected children. Due to the small family size of man these ideal ratios, first discovered by Mendel in the 19th century, are rarely found in individual families but can be approached when a large population is studied. Since the expression of a recessive trait requires the presence of two essentially identical genes, one from each parent, the finding of an increased rate of cousin marriage or other form of consanguinity among the parents of affected children is to be expected. If two people are related, there is an increased chance that they have inherited the same gene from one of their common ancestors. This increase in consanguinity is, of course, only noticeable in the case of rare recessives, since it would be extremely rare that two unrelated marriage partners carry the same rare gene. For example, if, as is approximately the case for galactosemia, one in 150 people carries one copy of the mutant gene, the chance of a mating between two such heterozygotes is 1:150×1:150, or 1:22,500. On the other hand, if an individual marries his first cousin, the chance that he or she will carry the same gene is 1:8, so the risk of both carrying the gene for galactosemia becomes 1:150×1:8, or 1:1,200. The risk of an affected child in the two situations is, respectively, one in 90,000 and one in 4,800, roughly a 19-fold difference. Of course, if one of the first cousins is a known carrier of the gene, the risk is markedly increased to one in 32.

The special case of genes carried on the X chromosome produces yet different pedigrees. It must be remembered that women, who have two X chromosomes, transmit one or the other of these to both their sons and daughters, while men, who have only one X chromosome, transmit this to all their daughters, but to none of their sons, who instead inherit their Y chromosome. It is immediately clear that there can be no male-to-male transmission of an X-linked gene. Most of these genes responsible for diseases (such as hemophilia A or Duchenne's muscular dystrophy) are expressed in females only if present in the homozygous state, that is, on both X chromosomes. They are, therefore, referred to as sex-linked recessives. However, the male carries only one X chromosome and, therefore, expresses the trait whenever he carries one copy of the mutant gene. This is the reason why the frequency of these diseases is so much higher among males than females. The frequency in males will be the same as the frequency of the mutant gene in all the X chromosomes of the population (in the case of hemophilia A about 1 in 10,000). Since the production of an affected female requires the chance meeting in the same fertilized egg of two mutant genes, one from an affected father and the other from usually heterozygous mother, the frequency of affected females is the square of the frequency in the population of the gene (in hemophilia A about $1:10,000^2$ or 1:100,000,000). Since male-to-male transmission is impossible, and since females do not express the disease when they carry only one copy of the mutant gene, the usual pedigree consists of an affected male with clinically normal parents and children, but with affected brothers, maternal uncles, and other maternal male relatives, as well as affected grandsons produced by their daughters.

It has been long appreciated that many normal traits, such as height, intelligence, and birth weight, have a significant genetic component, as well as a number of common diseases, such as diabetes mellitus, schizophrenia, hypertension, and coronary atherosclerosis. However, the pattern of inheritance of these traits does not follow the simple modes just described. Mathematical analysis of many of these has led to the conclusion that they follow the rules of polygenic inheritance, that is, are determined by a constellation of several genes, some derived from each parent.

Since each chromosome carries hundreds or thousands of genes in a linear order, chromosome maps of such linkage can be constructed. This is usually done by studying the coinheritance of specific alleles of two or more genes within families. Linkage, or the closeness of two genes in any of their allelic forms on the same chromosome, must be differentiated from association. The latter term is used when a specific allele of a gene is associated with a particular condition more frequently than by chance (eg, blood group O and peptic ulcer), and may imply something about the physiological function of the product of that allele.

Two more terms commonly used in genetics are penetrance and expressivity. Penetrance is a statistical term and indicates the proportion of individuals carrying a certain gene who can be detected. As our ability to detect the expression of a gene improves, the penetrance increases. Expressivity refers to the degree of expression of a gene in an individual. For example, full expressivity for osteogenesis imperfecta would include fragile bones, blue sclerae, and deafness. The presence of one or two of these findings comprises partial expressivity, while the absence of all three, occasionally found in carriers of this gene, is zero expressivity. The latter type of individual would fall into the group responsible for reduced penetrance. When we are able to detect the primary biochemical abnormality responsible for this disease, both zero expressivity and reduced penetrance will disappear.

Population genetics is also concerned with the distribution of genes in the human population. The frequency of various alleles, normal or abnormal, may vary from group to group. For example, the gene for sickle cell hemoglobin is found primarily in black populations deriving from western Africa and that for Tay-Sachs disease (hexosaminidase-A deficiency) primarily in Ashkenazi Jews, while the gene responsible for cystic fibrosis, so common in whites, is virtually absent in blacks.

It is clear that the information required to make a decision about the inheritance of a disease can only be obtained through a complete family history with attention to all the factors mentioned above. For example, a thorough search for consanguinity of the patient's parents is essential. The presence of an identical twin is often helpful, since both twins carry the same abnormal gene. However, it also should be clear that the absence of positive findings in the family history does not eliminate the possibility of an inherited disease, since new mutations, decreased penetrance, or undetected heterozygosity of the parents often mask genetic factors.

Cytogenetics

Human cytogenetics is concerned with the study of chromosomal abnormalities and their mechanisms. In order for such a study to be possible, it was necessary first to establish the normal human karyotype, that is, the normal number and structure of human chromosomes. Tissue culture techniques have permitted accurate counting and classification and the construction of karyotypes in which some pairs of chromosomes could be identified accurately and others grouped into classes by size and by the ratio of the lengths of the upper (short) and lower (long) arms. Recent developments have permitted accurate identification of each chromosome and its parts by special staining of repetitive DNA sequences, by affinity of quinacrine mustard or hydrochloride to certain segments (permitting visualization of these segments by fluorescence microscopy), by treatment with hot buffers before staining, and by the use of proteolytic enzymes. The last three techniques produce characteristic banding patterns for each chromosome, while the first two have been particularly useful for the identification of the Y chromosome.

The number of chromosomes in normal individuals is 46, of which 44 represent the 22 pairs of autosomes and the other two are the sex chromosomes. These are alike in the female (XX) and unlike in the male (XY).

The frequency of chromosome aberrations in the newborn is 1 in 200, while that in first trimester spontaneous abortions is about 50%. Thus, while cytogenetic disease represents a fairly common event in live births, the vast majority of chromosomal aberrations are lost in early fetal life.

The most commonly described syndromes found after birth are those due to abnormal numbers of chromosomes. These may involve the autosomes or the sex chromosomes. The autosomal syndromes are due to the presence of three instead of two of the relevant chromosomes. The most important of these trisomies are those of chromosome 21 (Down's syndrome or mongolism, incidence one in 650 live births), chromosome 13 (eye anomalies, cleft lip and palate, many other variable malformations, low birth weight, average life span under nine months, incidence one in 5,000) and chromosome 18 (hand anomalies, failure to thrive, micrognathia, low. birth weight, many other variable malformations, average life span under three months, incidence one in 5,000). Other autosomal trisomies possibly responsible for specific syndromes, but rarer in incidence, include those of chromosome 22 and one of the group of chromosomes including 6 to 12 (group C). Which of the latter group is responsible will be resolved with the use of the new techniques of identification. All of these trisomies increase in frequency with advancing maternal age, and the majority of cases appear to be due to meiotic nondisjunction. Normally, during formation of the ovum (or the sperm), the pairs of chromosomes separate so that only a single representative of each pair (that is, a total of 23) is found in each gamete. In meiotic nondisjunction, one of the pairs remains together, so that after fertilization there are three representatives of this chromosome in the new zygote forming the embryo instead of the pair found in normals. It is, of course, true that such nondisjunction also can produce a gamete lacking the particular chromosome, in which case fertilization with a normal gamete would produce a zygote with only one representative of the involved pair of chromosomes (monosomy), that is, a total of 45 chromosomes. No convincing examples of autosomal monosomy have been described in live births, and they are even rare in early abortion material, where the other trisomies, not found in live births, are found. This probably means that autosomal monosomy is lethal before implantation of the embryo.

Numerical aberrations of the sex chromosomes are quite frequent. The most important are the three with 47 chromosomes, that is, the normal 44 autosomes and sex chromosome constitutions of XXY, XYY, and XXX, and the only monosomy found in liveborns, that is, 45 chromosomes with only one sex chromosome, an X (often referred to as XO). The XXY constitution is found in the majority of patients with Klinefelter's syndrome (males with testicular dysgenesis, sterility and, in about 20%, gynecomastia, together with an increased incidence of mental retardation or other mental disease). Men with the XYY sex chromosome constitution are generally tall and have frequently been discovered among criminal populations. It is, however, not clear whether this represents a biological tendency toward criminality or simply reflects the greater incidence of tallness among criminals. Certainly plenty of normal males with XYY have been described, some of whom are fertile.

The XXX constitution is found in females who range in appearance from normal fertile women to some with characteristics resembling Turner's syndrome with an increased incidence of mental problems.

Women with an XO constitution (or 45,X, as they are designated by the currently accepted terminology) represent the majority of patients with Turner's syndrome (ovarian dysgenesis, short stature, sexual infantilism, gonadal streaks and sterility, with or without a variety of congenital defects, but generally without mental retardation). This chromosome abnormality is also frequently found among early spontaneous abortuses, and it appears that less than 5% come to term.

It is clear from the above that in man, as well as in other mammals, the presence of a Y chromosome determines male differentiation, no matter what the number of X chromosomes.

The number and type of sex chromosomes can, in most cases, be ascertained without actually preparing a karyotype. It appears that in female cells, one or the other of the two X chromosomes is completely or partially inactivated, that is, its genes are not expressed, and this inactive X can be seen in nondividing cells by special stains. The most commonly used cells for such a study are those of the buccal mucosa, a proportion of which show this X as the sex chromatin body (or Barr body) at the nuclear membrane. The presence of a Y chromosome can also be detected in nondividing cells with quinacrine due to the great affinity of this fluorescent compound for the Y chromosome. By studying the cells from individuals with differing numbers of X chromosomes, the rule has been established that the maximum number of Barr bodies found in cells of an individual is one less than the number of X chromosomes. Thus, XO females, XYY males, and XY males have no Barr body, XX females and XXY males have one, and XXX females have two. Males with an XYY constitution show two fluorescent Y bodies in their cells.

Some individuals, termed mosaics, have two or three cell lines in their bodies, differing from each other in chromosome number. Such mosaicism can be brought about by the loss of a chromosome during cell division (anaphase lag) or when both halves of a dividing chromosome end up in the same cell (mitotic non-disjunction). These errors occur after fertilization in one of the early mitotic divisions in the growth of the embryo. For the sex chromosomes, this can result in XX/XO (chromatin-positive Turner's or normal females), XY/XO (intersex, normal males, or Turner's), XXX/XO (XXX females with or without Turner's syndrome), XXX/XX and several others. Since the final appearance of the patient depends on the distribution of the various lines in the gonads, it is possible to see a patient with Turner's syndrome, who is sex-chromatin negative, with mosaicism of type XY/XO. Since the presence of the XY cell line in the gonad can lead to testicular differentiation, and since intraabdominal testicular tissue has a high malignant potential, it is important to study the actual chromosomes in order to select patients for laparotomy. Autosomal mosaicism can also occur and is particularly important in mongolism. Not only have some affected patients been found to have two cell lines, normal and trisomic 21, but apparently normal parents of such children have occasionally had this mosaicism. These parents are at very high risk for recurrence of mongolism in future offspring. It is, therefore, important to examine the chromosomes of parents of children with trisomy 21, expecially when the parents are in the younger age group.

Even more important from the point of view of genetic counseling are the structural chromosomal aberrations. The two major types of concern are translocations and deletions. Translocations arise as a result of breaks in two chromosomes followed by rejoining the fragments to the other broken chromosome. The rearranged products carry all the genes of the individual and the "balanced" carrier of two translocation chromosomes is clinically normal. However, during meiosis, the segregation of these chromosomes into the sperm or egg is quite variable, resulting after fertilization, in offspring who may be normal, clinically normal translocation carriers, "unbalanced" products who may have a variety of malformations and retardation, or may be aborted early in pregnancy. For example, if a woman carries a balanced translocation involving chromosome 21, about one third of her offspring will be chromosomally and clinically normal, one third will be clinically normal but carry the same balanced translocation as the mother, and one third will have mongolism. It is, therefore, most important to study the chromosomes of all children with mongolism in order to detect the families who are at unusually high risk for recurrence. Similarly, among families with multiple abortions, several malformed offspring, or a family history of malformations, a number have been found with translocations. The exact nature of these will be more accurately identifiable with the new techniques of chromosome identification.

Deletions, or chromosomes missing a piece, have also been associated with numerous malformations, such as the "cat-cry" syndrome due to a deletion of the short arm of chromosome 5. Since deletion can occasionally be due to familial translocations, family studies for proper counseling are necessary. The Philadelphia chromosome, containing a deletion in the short arm of a chromosome 22, is found in most cases of chronic myelocytic leukemia. While this is the only constant association between a specific chromosome abnormality and a malignant disease, most neoplasms show abnormal chromosomes of great variety. Similarly, patients with genetic diseases associated with a high incidence of neoplasia, such as Fanconi's anemia, show a tendency toward "spontaneous" chromosome breakage and rearrangement.

Biochemical Genetics

Our clearest understanding of genetic mechanisms in man have come from the study of human biochemical genetics. This field concerns itself with heritable variation in the structure or, in some cases, synthesis of proteins, primarily of enzymes. The protein that has so far given us the greatest amount of such information is hemoglobin. Adult hemoglobin (Hb A) consists of two pairs of chains, called α and β. Each of these is a chain of about 150 amino acids and carries a heme group with its iron for the purpose of oxygen transport. Pauling, in 1949, showed that sickle cell anemia is due to an alteration in hemoglobin, and Ingram, in 1957, demonstrated that this alteration is due to the substitution of a single amino acid in the chain. Since then, more than 100 such substitutions have been described in the two chains of Hb A. Some of these, like Hb S (the hemoglobin associated with sickle cell anemia) and Hb C have been associated with disease, but many others have no apparent deleterious effect and have been discovered by chance during the screening of normal populations. The method used for their detection is electrophoresis, a process that separates molecules on the basis of their electrical charge.

It is, therefore, possible to conclude that a mutational change in the DNA results in a substitution of an amino acid, as determined by the genetic code of the DNA. Since some mutations result in amino acid substitutions that do not alter the electric charge of the protein, and some, due to the existence of multiple code letters for single amino acids, do not even produce a substitution at all, the true variation in the genes for hemoglobin is even greater than that detected.

In addition, mutations exist that result in a reduction or complete lack of synthesis of one of the chains. These result in the various forms of thalassemia, of which Cooley's or Mediterranean anemia is one common example. Similar mechanisms may be operative in complete enzyme deficiencies and such conditions as analbuminemia.

Recent work, primarily by Harris, has demonstrated that as many as 25% of specific enzymes in man also show common variation (polymorphism), while, in addition, many other enzymes show rare genetic variants found in screening normal populations. Other variants, as in the hemoglobins, are associated with diseases generally called inborn errors of metabolism. These are frequently associated with enzyme variants that show markedly diminished enzyme activity and, therefore, cause disease when present in double dose, that is, in the homozygous state. Heterozygotes or carriers of these genes can frequently be detected because their cells show about one half the enzyme activity of that found in the cells from normal persons, but these carriers do not show symptoms of disease. For example, the parents of a child with deficiency of UDP-galactose transferase (galactosemia) or hexosaminidase A (Tay-Sachs disease) will have half the normal activity of the relevant enzyme, will show no signs of the disease but, since both carry a single copy of the mutant gene, will have a 25% risk for producing an affected child.

It is, therefore, clear that the mutations producing disease are only a small proportion of the enormous degree of genetic variation found in man and most other species. This biochemical individuality probably gives us the ability to adapt to changing environments as a population. That is, while some gene combinations may be unfavorable in certain environmental conditions, others may be favorable (as, for example, the gene for Hb S in an area endemic for falciparum malaria), and permit a portion of the population to survive and reproduce. This type of mechanism probably underlies Darwinian evolution, or "survival of the fittest."

Another important principle in the understanding of human genetic disease is that of heterogeneity. It has become evident that two patients with similar signs and symptoms may have these as the result of quite different mutant genes. For example, hereditary hemolytic anemia may be due to any of several inherited enzyme defects affecting red cell carbohydrate metabolism, abnormal hemoglobin, or membrane abnormalities. Each of these causes (at least 20) of hemolysis is determined by different genes. It is important to detect such heterogeneity for many reasons, not the least of which relates to genetic counseling. Two individuals carrying the same abnormal gene have a 25% risk of having affected offspring (if the trait is recessive), while two individuals carrying mutant genes representing different causes for a disease have essentially no risk of having children with this disease.

One subdivision of the field of human biochemical genetics is taking on increasing importance. This is pharmacogenetics, the study of inherited abnormal responses to drugs. The first example was that of unusual sensitivity to succinylcholine chloride, resulting in prolonged muscle paralysis and apnea, and due to one of several abnormalities in the enzyme plasma cholinesterase. Other such "idiosyncrasies" include inactivation of isoniazid, resistance to the anticoagulant action of warfarin sodium and others. It appears certain that further study will reveal that many other unusual responses to pharmacologic agents are based on inherited variations. When these are understood, it will be possible to test patients before treatment so as to avoid unanticipated drug reactions.

Immunogenetics

The field of immunogenetics concerns itself with two different topics. One is the study of genetic variation in antigenic determinants on cell surfaces, while the other deals with genetic aspects of normal and abnormal immune responses.

The first of these topics can again be subdivided as to cell type. One of the oldest and best studied aspects of human genetics concerns the blood groups found on red cells. Since the description in 1900 by Landsteiner of the ABO system, at least 14 additional genetic variations in blood groups have been discovered. These include, among others, MN, Rh, and a sex-linked group called Xga. Their importance in blood transfusion, erythroblastosis fetalis, and paternity testing is well known. Recent careful studies of the ABO system have demonstrated that while their specificity is determined by a series of sugars attached to a long carbohydrate chain in turn attached to a lipoprotein on the cell surface, their genetic determination is based on variation in specific enzymes, transglycosidases, responsible for attaching the appropriate sugar.

Erythroblastosis is caused by the immunization of an Rh-negative mother by her Rh-positive child, the positivity having been inherited from the father. Although not more than 20% of such incompatibilities result in significant immunization, it occurs frequently enough to be a cause of large numbers of fetal deaths or mental retardation produced by the severe jaundice in the perinatal time period. Since immunization almost always occurs at the time of delivery of the first incompatible offspring due to transplacental bleeds of fetal cells, it has become possible to prevent immunization and, therefore, disease of subsequent children by injecting anti-Rh antibody into the mother immediately after delivery of an incompatible child. This results in the inactivation of the Rh antigen on the fetal cells, preventing an antibody response by the mother.

Another recent major advance has been the discovery of the HL-A system of histocompatibility antigens on lymphocytes and other tissues. It is differences in these antigens that are responsible for rejection of transplanted tissues. The antigens are genetically determined by two closely linked genes, each having numerous different alleles spread through the population. Thus, while it is easy to find ABO compatible blood for transfusion, it is almost impossible to find HL-A compatible organs for transplantation, except from siblings. However, the ability to type individuals for the HL-A antigens on their lymphocytes has permitted more accurate matching of donor and host, resulting in great improvement in the retention of transplanted kidneys. This matching can be made even more accurate by the use of the mixed lymphocyte culture test, in which cocultivated lymphocytes, differing in their histo-compatibility types, stimulate each other to synthesize DNA and divide.

The second aspect of immunogenetics deals with the genetic deter-mination of gamma globulin antibody molecules, as well as with inherited abnormalities of antibody production and of the cellular immune response associated with delayed hypersensitivity. Normal synthesis of gamma globulins by lymphoid cells is a highly complex process, still not fully understood. The different classes of these molecules, such as γG, γM, and γA, each consist of two pairs of chains (analogous to those in hemoglobin) called light and heavy chains. Each of these is determined by its own genetic loci, those for heavy chains determining the class of immuno-globulin, while the light chains are capable of binding with the heavy chains of any class. Each chain consists of constant regions of amino acid sequences and a variable region in the area of the molecule defining antibody specificity. It is not clear at this time whether the constant and variable regions are specified by separate genes, and it is even less clear what are the genetic mechanisms behind the enormous variability within the variable portions responsible for the large variety in antibody specificity.

Inherited abnormalities of the immune response can affect gamma globulin production or cellular immunity or both. Abnormal individuals are generally detected because of their high susceptibility to infection, but the underlying reason can be one of numerous genetic variations. For example, hypogammaglobulinemia can occur in males in early childhood due to the presence of an X-linked gene or later in life in either sex due to one of several types of recessive genetic defects. Similar variations in onset, symptomatology, and mode of inheritance can underlie deficiencies in cellular immunity, such as is found in thymic alymphoplasia, or combined defects of immunoglobulin production and delayed hyper-sensitivity, such as in the Swiss type of agammaglobulinemia. In addition, high susceptibility to infection due to diminished host defense mechanisms can be due to a variety of nonimmunological inherited defects of leukocyte function. An example is the inability to kill intracellular bacteria by the cells of patients with the X-linked defect known as chronic granulomatous disease.

It is clear that host defense mechanisms are controlled by numerous genes, mutations of which can all lead to difficulties in combating environmental pathogens and, possibly, in defending the body from "foreign" cells typical of malignant tumors.

Somatic Cell Genetics

Recent improvements in techniques of cell culture have permitted not only the great increase of diagnostic genetic studies of human disease but have made possible new approaches to the study of human genetics in the test tube. Diagnostic techniques include the cytogenetic studies already described, with the use of peripheral blood lymphocytes, bone marrow, and fibroblast cultures developed from skin biopsies. The fibroblasts and lymphocytes can also be used for the diagnosis of inborn errors of metabolism as well as the detection of heterozygous carriers for these disorders.

Cells cultured from amniotic fluid obtained in the second trimester of pregnancy by amniocentesis of mothers at risk for having an abnormal child are being used for cytogenetic and biochemical diagnostic studies of the fetus.

The new technique for developing mass cultures of permanent lymphoid cell lines from any donor will allow careful study of the abnormal enzymes found in genetic disease, their effects on cellular metabolism, and will permit attempts at correction of the biochemical defect by exposure of the cells to normal enzymes. These cells are also useful for the study of attempts at correcting the basic genetic defect by introducing new genes coding for the missing or abnormal enzyme. This could be achieved by the use of human DNA or microbial genes carried into the cells by viruses. Such studies, if successful, could eventually lead to therapy of currently lethal disorders.

It is now possible to fuse human cultured cells with those of other species. Such hybrids tend to lose human chromosomes. By characterizing the karyotype and enzymological make up of the hybrid cells, one can discover the localization of genes coding for specific enzymes on specific chromosomes. This technique will be one of the most useful in constructing the human genetic map, that is, the assignment of genes and their order on the individual chromosomes.

The exposure of cultured cells to environmental agents, followed by a search for chromosomal damage and evidence of biochemical changes resulting from mutation, at last provides us with a method for monitoring chemicals for mutagenic effects. Until now all such studies have been restricted to microbes or experimental animals, but the use of human cells will allow more relevant conclusions.

Clinical Genetics and Counseling

It cannot be stressed strongly enough that the pivotal person in clinical genetics is the physician. Unless the diagnosis is correct, all other efforts are of little use and frequently can be misleading or harmful. The clinical geneticist must, therefore, not only be acquainted with most of the great variety of genetic syndromes, but must have available diagnostic laboratories capable of appropriate cytogenetic and biochemical studies, as well as the cooperation of consultants in the subspecialties pertinent for each case. This is, of course, even more essential for the practicing physician who will see each of the disorders only rarely, although he may see a sizable total number of patients with genetic problems. If there is any doubt in his mind about the accuracy of his diagnosis, especially in view of the recognition of multiple causation of similar symptoms (heterogeneity), he must consult an experienced human geneticist before attempting to give advice and counseling. It can be fatally damaging to a family to be told that they have a high risk of an abnormal offspring when their chances may be almost as good as that of the general population. The reverse can be just as bad when the second abnormal child arrives in the face of optimistic statements.

It is also important to keep in mind that a diagnosis of genetic disease will usually only be made if the physician develops a high index of suspicion. It is, therefore, necessary to obtain a family history routinely and that this history include several generations and such data as parental consanguinity. It is frequently also crucial to perform careful examinations on first degree relatives in order to look for minor signs of inherited disorders with variable expressivity. For the purpose of counseling the various members of a family, heterozygote testing or chromosome analysis of healthy individuals often becomes necessary.

A few genetic diseases are now amenable to therapy, such as the dietary restrictions of phenylketonuria and galactosemia. It is hoped that new therapeutic techniques will become available for more disorders, but until they do, genetic counseling remains the major method for the prevention of abnormal offspring. Much of genetic counseling still depends on informing a family of the statistical risk of having an abnormal child. However, for an increasing number of disorders, the new development of prenatal diagnosis has provided the counselor with a powerful tool enabling him to give the family exact information as to the status of their unborn child.

Amniocentesis, or the obtaining of a small sample of amniotic fluid through a transabdominal needle, is done between the 14th and 18th week after the last menstrual period. The cells in this fluid are fetal in origin and, therefore, represent the fetal chromosomal and genetic complement. These cells can be cultured and their chromosomes examined after 10 to 20 days. After another week, sufficient cellular growth occurs so that a variety of enzymes can be assayed in the cell homogenates. By those techniques a number of genetic diagnoses can be made in fetuses of mothers at risk, and the information becomes available in time to perform a therapeutic abortion if the fetus is affected and, of course, if this is the parents' decision. Here, as in all aspects of genetic counseling, the bias of the counselor must not be permitted to influence the decision of the family. It is the counselor's duty to provide them with all the available information and its meaning, as well as to answer any questions that can be answered. After that, with the necessary psychological support of the counselor, the family must arrive at the decision of what to do, whether this relates to having further children or obtaining an abortion.

By means of examining amniotic fluid cell cultures, it has already become possible to diagnose the sex of the fetus in sex-linked disorders, such as Duchenne's muscular dystrophy and hemophilia, in which only males are affected. In addition, chromosomal analysis in older mothers or in families carrying translocations has allowed the prenatal diagnosis of mongolism and other chromosome defects. Finally, by performing the appropriate enzyme assays, geneticists have detected fetuses with about 20 different inborn errors of metabolism, and the list of possible diagnoses is rapidly expanding.

With the constant advances in genetic diagnosis, treatment, and preventive counseling, the bleak prospects of 20 years ago look much brighter. With the reduction of infectious and, to some extent, degenerative disease, a larger proportion of hospital beds and outpatient visits are accounted for by genetic disorders. Recognition of these disorders by the physician, followed by appropriate family studies and counseling will do as much or more to reduce their frequency as will the necessary continuation of research.

This investigation was supported by Career Scientist Award I-513 from the New York City Health Research Council.

Recommended Reading and References

Brief Reviews

McKusick VA: *Human Gentics,* ed 2. Englewood Cliffs, NJ, Prentice-Hall Inc, 1969.

Redding A, Hirschhorn K: *Guide to Human Chromosome Defects, Original Article Series.* New York, National Foundation–March of Dimes, 1968, vol 4.

Detailed Books

Hamerton JL: *Human Cytogenetics.* New York, Academic Press, 1971, vol 1 and 2.

Harris H: *The Principles of Human Biochemical Genetics.* Amsterdam, North-Holland Pub Co, New York, American Elsevier Pub Co, 1970.

Bodmer WF, Cavalli-Sforza LL: *The Genetics of Human Populations.* San Francisco, WH Freeman & Co Publishers, 1971.

Annuals

Harris H, Hirschhorn K (eds): *Advances in Human Genetics.* New York, Plenum Press.

Steinberg AG, Bearn AG (eds): *Progess in Medical Genetics.* New York, Grune & Stratton.

Reference Books

McKusick VA: *Mendelian Inheritance in Man,* ed 3. Baltimore, Johns Hopkins Press, 1971.

Genetic Services, International Directory, ed 3. New York, National Foundation–March of Dimes, 1971.

3
Human Chromosome Abnormalities as Related to Physical and Mental Dysfunction

JOHN H. HELLER

The relationship of human disease syndromes to chromosome aberrations is assuming an increasingly greater role in the detection, diagnosis, treatment and prediction of mental and physical defects in man. By means of karyotype analysis one is enabled to recognize previously unknown syndromes and to differentiate between separate but phenotypically similar entities. Proper diagnosis permits suitable therapeutic measures to be undertaken and enables genetic counselors to assess correct risks in many instances. Recent refinements in sampling embryonic cells by amniocentesis make it feasible to determine, in high risk cases, whether the embryo has a chromosome abnormality or whether it is a male, which has a high risk of sex-linked genetic defect. Termination of pregnancy can be recommended on the basis of this knowledge.

Classes of Chromosome Abnormalities

Chromosome abnormalities have been known in plant and animal species for a very long time. They occur firstly as variations in the number of chromosomes per cell deviating from the normal two sets (maternal and paternal), existing either as complete multiples of sets, a condition called polyploidy (triploidy, tetraploidy, etc.), or as addition or loss of chromosomes within a set, a situation known as aneuploidy (monosomy, trisomy, tetrasomy, etc.). The origin of deviations in chromosome number is known to be through nondisjunction, either during the meiotic divisions in the maturation of the germ cells or during mitotic divisions in the developing individual, or through lagging of chromosomes at anaphase of cell division.

Secondly, chromosome aberrations occur as structural modifications such as duplications, deficiencies, translocations, inversions, isochromosomes, ring chromosomes, etc. These aberrations result from chromosome breakage and reunion in various patterns different from the normal sequence of loci. In most cases, especially the "spontaneous" instances, the cause of chromosome breaks is unknown, but many extraneous agents have been demonstrated experimentally to be efficacious in inducing fragmentation. Foremost among these agents is ionizing radiation but many chemical substances (alkylating agents, nitrosocompounds, antibiotics, DNA precursors, etc.) and viruses have been implicated.

Genetic Effects of Chromosome Aberrations

The striking genetic alterations accompanying chromosome aberrations were brilliantly analyzed by Blakeslee and coworkers on *Datura,* and by the *Drosophila* workers (Morgan, Bridges, Muller, Sturtevant, Painter, Patterson and many others). The task was greatly facilitated in *Drosophila* by the fortunate circumstance in the larval salivary glands where the giant polytene chromosomes exhibit intimate somatic pairing as well as characteristic banding patterns that permit identification of specific gene loci.

Particularly illuminating were Bridges' analyses of sex chromosomes and sex determination in *Drosophila,* utilizing the phenomenon of nondisjunction of the sex chromosomes and culminating in the genic balance theory of sex determination. In this insect the female normally has two X chromosomes plus the autosomes, and the male has one X and one Y. Two X chromosomes and one Y chromosome results in a female, whereas a chromosome constitution of XO produces a sterile male.

In contrast, the Y chromosome in mammals has a strongly masculinizing influence. The presence of a single Y is sufficient to induce differentiation into a male phenotype in the presence of one to five X chromosomes. The XO consitution differentiates into a female phenotype in both mouse and man.

Mammalian Chromosome Studies

The first reported instance of chromosome aberration in mammals was discovered by genetic methods in the waltzing mouse by William H. Gates in 1927[37] and analyzed cytologically by T. S. Painter[68]. Many difficulties in techniques prevented accurate counting and analysis of mammalian chromosomes—large number and relative small size of chromosomes, tendency to clump on fixation, cutting of chromosomes in sectioned material, etc. Even the somatic chromosome number in man was accepted erroneously as 48 until 1956 when Tjio and Levan[88] established the correct count of 46. This count was quickly confirmed by Ford and Hammerton[28] and in 1959 the first positive correlation of a chromosome abnormality and human disease syndrome was made by Lejeune et al.[54] (also Jacobs et al.[44]) —the trisomic number 21 chromosome, and Down's syndrome or mongolism. Shortly thereafter Klinefelter's[46] and Turner's[30] syndromes were identified with XXY and XO sex chromosome constitutions respectively, and in rapid succession reports of many other

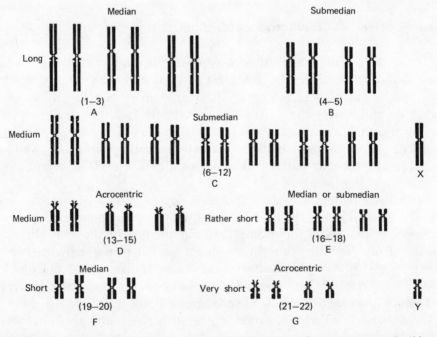

FIGURE 1. Idiogram of normal male with 22 pairs of autosomes and XY sex chromosome constitution (modified from Patau,[69] Sohval,[84] Ferguson-Smith et al.,[27] and Palmer and Funderburk[68a]).

human chromosome abnormalities appeared, such as trisomy 17, trisomy 18, partial trisomy, ring X chromosome, sex chromosome mosaics, cri du-chat syndrome, et cetera[9, 26]

This sudden explosion of human chromosome studies, in contrast to the long delay of confirmation in human cells of chromosome abnormalities long known in plants and other animals was made possible by new techniques of preparation. The accumulation of many cells in the metaphase stage of mitosis with colchicine, the use of hypotonic solution to swell the cells and separate chromosomes on the spindle, the discovery that phytohemagglutinin stimulates mammalian peripheral lymphocytes to undergo mitosis, and the method of squashing or spreading on slides of loose cells taken from bone marrow or tissue culture, all contributed to the rapid and accurate analysis of mammalian and human chromosome number and structure.

Karyotype analysis involves the careful comparison of chromosomes in a particular individual to the standard pattern for human cells, including precise measurements of lengths, arm ratios and other morphological features. Special attention is given to comparison of homologous chromosomes where differences may indicate abnormalities. An idiogram is a diagrammatic representation of the entire standard chromosome complement, showing their relative lengths, position of centromeres, arm ratios, satellites, secondary constrictions and other features. Figure 1 shows an idiogram of a normal human male with 22 pairs of autosomes and XY sex chromosome constitution. A karyotype is constructed from photographs of chromosomes which are arranged in pairs similar to the idiogram. Figure 2 shows a karyotype of a normal human female.

Incidence of Human Chromosome Anomalies

Chromosome anomalies are relatively frequent events. They have been estimated to occur in 0.48 percent of all newborn infants (one in 208)[81]. At least 25 percent of all spontaneous miscarriages result from gross chromosomal errors[13]. The general incidence of chromosome abnormalities in abortuses is more than fifty times the incidence at birth.

Although it is impossible to obtain an accurate total of victims suffering from effects of chromosome aberrations, one can make rough calculations on the basis of their estimated frequencies in the population of the United States assuming that there is no appreciable difference in life expectancy

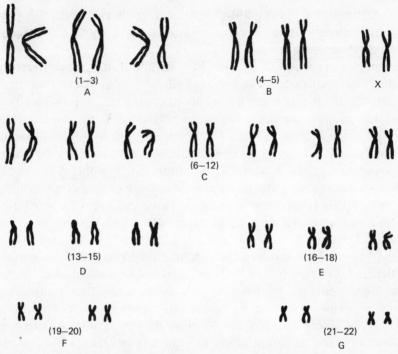

FIGURE 2. Karyotype of a normal human female with 22 pairs of autosomes and two X chromosomes.

between these individuals and those with normal chromosome complements. Although this assumption probably is unjustifiable, it suffices for this rough calculation. Among the current population of 202 million we arrive at a figure of 1,136,971 total afflicted with chromosome abnormalities. This total probably represents an underestimate since it does not include all types of chromosome aberrations. Table I indicates totals for a number of specific syndromes.

Syndromes Related to Autosome Abnormalities

Down's Syndrome

This defect results from duplication of all or part of autosome 21, either in the trisomic state or as a translocation to another chromosome, usually a 13-15 (D group) or 16-18 (E group) but may be to another G group chromosome. The overall incidence is about 1 in 700 live births[71], but the trisomic type is correlated with age of the mother, having a

TABLE I. Total frequencies in the United States of various types of chromosomal abnormalities, calculated on the basis of 202 million current population and the estimated frequency of each abnormality. (It must be noted that the grand total does not include all types of chromosome aberrations, therefore must be lower than the real value)

Syndrome	Chromosome Number	Estimated Incidence	Calculated Number in U.S.
Down's trisomy 21	47	1 in 700	288,571
Trisomy D	47	1 in 10,000	20,200
Trisomy E	47	1 in 4000	50,500
Trisomy X	47	1 in 10,000 females	101,000
Turner's XO	45	1 in 5000 females	20,200
Klinefelter's XXY	47	1 in 400 males	252,500
Double Y XYY	47	1 in 250 males	404,000
		Total	1,136,971

frequency of about 1 in 2000 in mothers under 30 years of age, and increasing to 1 in 40 in mothers aged 45 or over. The translocation type constitutes about 3.6 percent of cases and is unrelated to the mother's age, but is transmitted in a predictable manner. Among mental retardates mongoloids represent 16.7 percent.

Clinical features include physical peculiarities ranging from slight anomalies to severe malformations in almost every tissue of the body. Typical appearance of a mongoloid shows slanting eyes, saddle nose, often a large ridged tongue that rolls over a protruding lip, a broad, short skull and thick, short hands, feet and trunk. Frequent complications occur: cataract or crossed eyes, congenital heart trouble, hernias, and a marked susceptibility to respiratory infections. They exhibit characteristic dermatoglyphic patterns on palms and soles. Also they have many biochemical deviations from normal, such as decreased blood-calcium levels and diminished excretion of tryptophane metabolites. Early ageing is common.

All mongoloids are mentally retarded; they usually are 3 to 7 years old mentally. Among the relatively intelligent patients, abstract reasoning is exceptionally retarded.

Female mongoloids are fertile and recorded pregnancies have yielded approximately 50 percent mongoloid offspring. Fortunately male mongoloids are sterile. Examination of their testes reveals varying degrees of spermatogenic arrest correlated with the abnormal chromosome features.

Among mongoloids there is a prevalence of leukemia in childhood: the incidence is some twenty times greater than in the general population.

Simultaneous occurrence with other syndromes such as Klinefelter's also is found, and many cases of mosaicism have been described.

E trisomy Syndrome

This is another autosomal anomaly, which involves chromosomes 16, 17 and 18, and is estimated to occur at a frequency of 1 in 4000 live births[20]. Many others die before birth, thus contributing to the large number of miscarriages and stillbirths. These individuals survive only a short time, from one-half day to 1460 days, with an average of 239 days, but females live significantly longer than males.

Trisomy 17 Syndrome

Many serious defects usually are present in afflicted individuals[25]: odd shaped skulls, low-set and malformed ears, triangular mouth with receding chin, webbing of neck, shield-like chest, short stubby fingers, and toes with short nails, webbing of toes, ventricular septal defect and mental retardation, as well as abnormal facies, micrognathia and high arched palate.

Trisomy 18 Syndrome

This anomoly[70,82] is characterized by multiple congenital defects of which the most prominent clinical features are: mental retardation with moderate hypertonicity, low-set malformed ears, small mandible, flexion of fingers with the index finger overlying the third, and severe failure to thrive. It generally results in death in early infancy. Its frequency increases with advanced maternal age. Three times as many females as males have been observed; one would expect that more males with this syndrome will be found among stillbirths and fetal deaths.

D Syndrome

This trisomy[19,56,70,83] involves chromosomes 13, 14 and 15, and has an estimated frequency of about 1 in 10,000 live births. Many others die in utero. Survival time has been reported from 0 to 1000 days, with an average of 131 days.

Clinical features include: microcephaly, eye anomalies (corneal opacities, colobomata, microphthalmia, anophthalmia), cleft lip, cleft palate, brain anomalies (particularly arrhinencephaly), supernumerary digits, renal anomalies (especially cortical microcysts), and heart anomalies.

Trisomy 22 Syndrome

This syndrome produces mentally retarded, schizoid individuals. Reports of its occurrence are too few to permit an estimate of its frequency in the population.

Cri-du-chat Syndrome

Lejeune *et al.*[55] first described this anomaly in 1965, which involves a deficiency of the short arm of a B group chromosome, number 5. Translocations appear to be a common cause of the defect, an estimated 13 percent of cases being associated with translocations; described cases have had B/C, B/G, and B/D translocations[23]. The high proportion emphasizes the importance ·of unbalanced gamete formation in translocation heterozygotes as a cause of this syndrome. Among parents the frequency of male and female carriers is approximately the same, a situation that contrasts with the much greater frequency of female carriers of a D/G translocation among parents of translocation mongoloids.

Typical clinical features of cri-du-chat individuals are: low birth weight, severe mental retardation, microcephaly, hypertelorism, retrognathism, downward slanting eyes, epicanthal folds, divergent strabismus, growth retardation, narrow ear canals, pes planus and short metacarpals and metatarsals. About 25 to 30 percent of them have congenital heart disorders. A characteristic cat-like cry in infancy is responsible for the name of the syndrome. The cry is due to a small epiglottis and larynx and an atrophic vestibule. However, this major diagnostic sign disappears after infancy, making identification of older cases difficult.

An estimate of the frequency of this syndrome is given as over 1 percent but less than 10 percent of the severely mentally retarded patients. Many have IQ scores below 10, and most are institutionalized.

Philadelphia Chromosome

Finally, among autosomal aberrations, a deleted chromosome 21 occurs in blood-forming stem cells in red bone marrow. This deletion, which shows up long after birth, appears to be the primary event causing chronic granulocytic leukemia. This aberration was discussed in 1960 by Nowell and Hungerford[67] (also Baikie *et al.*[3])

Syndromes Related to Sex Chromosome Aberrations

The great majority of known chromosomal abnormalities in man involve the sex chromosomes. In one survey (that excluded XYY) it was estimated that abnormalities occurred in 1 out of every 450 births; if the recent estimate of XYY[81] is correct, the frequency actually is much higher. Increased knowledge about sex chromosome aberrations is probably related to the greater concentration of attention on patients with sexual disorders, but is due in part to the ability to detect carriers of an extra X chromosome by the so-called sex chromatin body or Barr body[6]. This structure is a stainable granule at the periphery of a resting nucleus and, according to the Lyon hypothesis[57], is considered to be an inactivated X chromosome. A normal female cell had one Barr body, since it has two X chromosomes, and is said to be sex chromatin positive (or one positive). A normal male cell has no Barr body and is said to be sex chromatin negative.

Klinefelter's Syndrome

The first sex chromosome anomaly described in 1959 by Jacobs and Strong[46] and also by Ford et al.[29] was the XXY constitution that is typical of Klinefelter's syndrome. Buccal smears from these patients are sex chromatin positive. They can be tentatively diagnosed by this test along with clinical symptoms. Final confirmation of diagnosis can be achieved by karyotype analysis using either bone marrow aspiration or peripheral blood culture.

Victims of Klinefelter's syndrome are always male but they are generally underdeveloped, eunachoid in build, with small external genitalia, very small testes and prostate glands, with underdevelopment of hair on the body, pubic hair and facial hair, frequently with enlarged breasts (gynecomastia), and many have a low IQ.

The classical type with two X chromosomes and one Y chromosome was the first case discovered, but subsequently chromosome compositions of XXXY, XXXXY, XXYY[66] and XXXYY[7,8,63,77] have been reported. In addition, numerous mosaics have been described, including double, triple and quadruple numeric mosaics, as well as combinations of numeric and structural mosaics. These conditions are summarized in Table II. They all resemble the XXY Klinefelter's phenotypically and are considered modified Klinefelter's syndromes. The classical XXY type may have low normal mental development or may be retarded, but other types show increasingly greater mental retardation.

The incidence of Klinefelter's syndrome is estimated to be 1 in 400 mole live births, which represents from 1 to 3 percent of mentally deficient patients. This condition also has been correlated with age of the mother: the older the mother, the greater the risk of having such a child. These individuals usually are sterile. Spermatogensis is generally totally absent. Hyalinization of the semeniferous tubules begins shortly before puberty. Congenital malformations are rare. Mental retardation is present in approximately 25 percent of affected individuals, and mental illness may be more common than in the general population.

TABLE II. Reported Sex Chromosomal Constitutions in Klinefelter's syndrome (modified from Reitalu[77])

		Sex Chromosomal Constitution		
Only one karyotype observed per individual		XXY		
		XXYY		
		XXXY		
		XXXYY		
		XXXXY		
Numeric mosaics	Double	XX	XXY	
		XY	XXY	
		XY	XXXY	
		XXY	XXYY	
		XXXY	XXXXY	
		XXXX	XXXXY	
	Triple	XY	XXY	XXYY
		XX	XXY	XXXY
		XY	XXY	XXXY
		XO	XY	XXY
		XX	XY	XXY
		XXXY	XXXXY	XXXXYY
		XXXY	XXXXY	XXXXXY
	Quadruple	XXY	XY	XX	XO
Numeric and structural mosaics	Double	XXY	XXxY	
	Triple	XY	XXY	XXxY
		XxY	Xx	XY

Turner's Syndrome

Female gonadal dysgenesis was described by Turner in 1938 as a syndrome of primary amenorrhea, webbing of the neck, cubitas valgus and short stature, coarctation of aorta, failure of ovarian development and hormonal abnormalities. Patients exhibit sexual infantilism; their breasts are usually underdeveloped, nipples often widely spaced, particularly in those subjects who have a shield or funnel chest deformity. Usually sexual hair is scanty; external genitalia are infantile; labia small or unapparent; clitoris usually normal, although may be enlarged. The uterus is infantile; the tubes long and narrow; the gonads represented by long, narrow, white streaks of connective tissue in normal position of ovary. They are almost always sterile. Hormonal secretions usually are abnormal. Shortness of stature is characteristic and many other skeletal abnormalities occur. Peculiar facies include small mandible, anti-mongolian slant of eyes, depressed corners of mouth, low-set ears, auricles sometimes deformed.

Cardiovascular defects are frequent, the most common being coarctation of the aorta. Slight intellectual impairment is found in some patients, particularly those with webbing of the neck.

In 1954 it was discovered that many patients with ovarian agenesis were sex chromatin negative, and in 1959 Ford and colleagues[30] gave the first chromosome analysis showing that Turner's syndrome has the sex chromosome abnormality of only one X chromosome (XO) rather than two X's. It was quickly confirmed by Jacobs and Keay[4] and by Fraccaro *et al*[33].

Mosaicism is known to exist—both 45 chromosome cells and 46 chromosome cells occur side by side in tissues of the individual—and can result from non-disjunction in early embryonic development. Isochromosomes sometimes are involved, e.g., creating a situation with 3 long arms of the X chromosome but only 1 short arm.

The incidence of XO Turner's syndrome is estimated as 1 in approximately 5000 women; many die in utero.

Large scale screening of newborn babies by buccal smears can permit detection of chromatin negative females, chromatin positive males, and double, triple, quadruple and quintuple positive cases of either sex[58]. Table III shows the relationship between sex chromosome complements and sex chromatin pattern.

Triplo-X Syndrome

Females containing three[47], four and five X chromosomes are known[4,47,63]. The triplo-X syndrome is thought to have an incidence of about 1 in 800 live female births. This syndrome was first described by Jacobs *et al.*[47] in 1959. Although it has no distinctive clinical nature, menstrual irregularities may be present, secondary amenorrhea or premature menopause. Most cases have no sexual abnormalities and many are known to have children. Thos most characteristic feature of 3X females is mental retardation. Quadruple-[14] and quintuple-X[50] syndrome are much rarer. These individuals are mentally retarded, usually the more X chromosomes present, the more severe the retardation. Frequently these individuals are fertile.

An extra X chromosome confers twice the usual risk of being admitted to a hospital with some form of mental illness. The loss of an X, on the other hand, has no association with mental illness; thus the chance of mental hospital admission is not raised for an XO female. An extra X chromosome also predisposes to mental subnormality. The prevalence of psychosis among patients in hospitals for the subnormal is unusually high in males with two or more X chromosomes.

TABLE III. The Relationship Between Sex, Sex Chromosome Complement and Sex Chromatin Pattern (Modified from Miller[63])

	Sexual Phenotype	
Sex Chromatin Pattern	Female	Male
–	XO	XY
–	XY	XYY
	(testicular feminization)	
+	XX	XXY
		XXYY
++	XXX	XXXY
		XXXYY
+++	XXXX	XXXXY
++++	XXXXX	

Numerous other sex chromosome anomalies occur[38], many involving mosaics and structural chromosome aberrations. For example, occasionally an XY embryo will differentiate into a female, a situation referred to as testicular feminization male pseudohermaphrodite (Morris syndrome)[60]. These individuals have only streak gonads and vestigial internal genital organs. They usually have undeveloped breasts and do not menstruate. They are invariably sterile[76].

Still other sexual abnormalities are intersexes and true hermaphrodites, many of which have an XX sex chromosome constitution or are mosaics for sex chromosomes such as XO/XY or XX/XY or more complicated mixtures[31]. Sex chromosome mosaicism is very common. Almost every sex chromosome combination found alone has been found in association with one or more cell lines with a different sex chromosome constitution. These mosaics exhibit quite a variable expression; for example in an XO/XY mosaic the external genitalia can appear female, male or intersexual[85].

The YY Syndrome

The male with an extra Y chromosome (XYY) has attracted much attention in the public press as well as in scientific circles because of his reputed antisocial, agressive and criminal tendencies[1,2,64]. Although this abnormality belongs in the above category of syndromes related to sex chromosome aberrations, it has been singled out for special discussion because of its social and legal implications.

Evidence supporting the existence of a double Y syndrome has accumulated within the last six years. Studies in Sweden[32] showed an unusually large number of XXYY and XYY men among hard-to-manage patients in mental hospitals. These observations received impressive confirmation in studies of maximum security prisons and hospitals for the criminally insane in Scotland where an astonishingly high frequency (2.9 percent) of XYY males were found[48]. This was over fifty times higher than the then current estimate of 1 in 2000 in the general population. Subsequently many additional studies on the YY syndrome have appeared and a compsite picture of the XYY male emerged[5,19,15,21a,34,35,41,72-75,78,80,92].

The principal features of the extra Y syndrome appear to be exceptional height and a serious personality disorder leading to behavioral disturbances. It seems likely it is the behavior disorder rather than their intellectual incompetence that prevents them from functioning adequately in society[18].

Clinically the XYY males are invariable tall (usually six feet or over) and frequently of below-average intelligence. They are likely to have unusual sexual tastes, often including homosexuality. A history of antisocial behavior, violence and conflict with the police and educational authorities from early years is characteristic[86] of the syndrome.

Although these males usually do not exhibit obvious physical abnormalities[12,24,40,42,52,91], several cases of hypogonadism[11], some with undescended testes, have been reported. Others have epilepsy, malocclusion and arrested development[87], but these symptoms may be fortuitously associated. One case was associated with trisomy 21[61], another with pseudohermaphrodism[35]. The common feature of an acne-scareed face may be related to altered hormone production. The criminally agressive group were found to have evidence of an increased androgenic steroid production as reflected by high plasma and urinary testosterone levels[12,43]. If the high level of plasma testosterone is characteristic of XYY individuals it suggests a mechanism through which this condition may produce behavioral changes, possibly arising at puberty.

Antisocial and aggressive behavior in XYY individuals may appear early in life, however, as evidenced by a case reported by Cowie and Kahn[22]. A prepubertal boy with normal intelligence, at the age of 4½ years, was unmanageable, destructive, mischievous and defiant, overadventurous and without fear. His moods alternated; there were sudden periods of overactivity at irregular intervals when he would pursue his particular antisocial activity with grim intent. Between episodes he appeared happy and constructive. The boy was over the 97th percentile in height for his age, a fact that supports the view that increased height in the XYY syndrome is apparent before puberty.

It has been suggested that the ordinary degree of aggressiveness of a normal XY male is derived from his Y chromosome, and that by adding another Y a double dose of those potencies may facilitate the development of aggressive behavior[65] under certain conditions. A triple dose (XYYY) would be present in the case reported by Townes et al.[90].

The first reported case of an XYY constitution[39,79] was studied because the patient had several abnormal children, although he appeared to be normal himself. Until recently, reports of the XYY constitution have been uncommon, probably because no simple method exists for screening the double-Y condition that is comparable to the buccal smear—sex chromatin body technique for detecting an extra X chromosome. Another possible explanation for the rarity of reports on the XYY karyotype is the absence of a specific phenotype in connection with it. Most syndromes with a chromosome abnormality are ascertained because of some symptom or clinical sign that indicates a need for chromosome analysis. Consequently there have been few studies that place the incidence of this chromosome abnormality in its proper perspective to the population as a whole.

Very recently a study of the karyotypes of 2159 infants born in one year was made by Sergovich et al.[81]. These investigators detected 0.48 percent of gross chromosome abnormalities. In this sample the XYY condition appeared in the order of 1 in 250 males, which would make it the most common form of aneuploidy known for man. The previous estimate was about 1 in 2000 males. If this figure of 1/250 is valid for the population as a whole, it means that the great majority of cases go undetected and consequently must be phenotypically normal and behave near enough to the norm to go unrecognized.

Several cases of asymptomatic males have been published, including the first one described (Sandberg et al.[79] and Hauschka et al.[39]), which proved to be fertile. It appears that the sons of XYY men do not inherit their father's extra Y chromosome[59a].

Another fertile XYY male, reported by Leff and Scott[53], had inferiority feelings, was slightly hypochondriacal and obsessional, and not very aggressive. He gave a general impression of emotional immaturity. He was 6 feet, 6 inches tall, healthy, with normal genitalia and electroencephalogram. His IQ was 118. Wiener and Sutherland[93] discovered by chance an XYY male who was normal; he was 5 feet, 9½ inches tall, with normal genitalia and body hair, normal brain waves, and with an IQ of 97. He exhibited a cheerful disposition and mild temperment, had no apparent behavioral disturbance and never required psychiatric advice. This case supports the idea that an XYY male can lead a normal life.

Social and Legal Implications of the YY Syndrome

The concept that when a human male receives an extra Y chromosome it may have an important and potentially antisocial effect upon his behavior is supported by impressive evidence[15,21]. Lejuene states that "There are no born criminals but persons with the XYY defect have considerably higher chances." Price and Whatmore[74] describe these males as psychopaths, "unstable and immature, without feeling or remorse, unable to construct adequate personal relationships, showing a tendency to abscond from institutions and committing apparently motiveless crimes, mostly against property." Casey and coworkers[16] examined the chromosome complements in males 6 feet and over in height and found: 12 XYY among 50 mentally subnormal and 4XYY among mentally ill patients detained because of antisocial behavior; also 2 XYY among 24 criminals of normal intelligence. They concluded that their results indicate that an extra Y chromosome plays a part in antisocial behavior even in the absence of mental subnormality. The idea that criminals are degenerates because of bad heredity has had wide appeal. There is no doubt that genes do influence to some extent the development of behavior. The influence may be strongly manifested in some cases but not in others. Some individuals appear to be driven to aggressive behavior.

Several spectacular crime cases served to publicize this genetic syndrome, and it has been played up in newspapers, news magazines, radio and television. In 1965 Daniel Hugon, a stablehand, was charged with the murder of a prostitute in a cheap Paris hotel. Following his attempted suicide he was found to have an XYY sex chromosome constitution. Hugon surrendered to the police and his lawyers contended that he was unfit to stand trial because of his genetic abnormality. The prosecution asked for five to ten years; the jury decided to give him seven.

Richard Speck, the convicted murderer of eight nurses in Chicago in 1966, was found to have an XYY sex chromosome constitution. He has all the characteristics of this syndrome found in the Scottish survey: he is 6 feet 2 inches tall, mentally dull, being semiliterate with an IQ of 85, the equivalent of a 13-year-old boy. Speck's face is deeply pitted with acne scars. He has a history of violent acts against women. His aggressive behavior is attested by his record of over 40 arrests. Speck was sentenced to death but the execution has been held up pending an appeal of the conviction.

In Melbourne, Australia, Lawrence Edward Hannell, a 21-year-old laborer on trial for the stabbing of a 77-year-old widow, faced a maximum sentence of death. He was found to have an XYY constitution, mental retardation, an aberrant brain wave pattern, and a neurological disorder. Hannel pleaded not guilty by reason of insanity, and a criminal court jury found him not guilty on the ground that he was insane at the time of the crime. A second Melbourne criminal with an XYY constitution, Robert Peter Tait, bludgeoned to death an 81-year-old woman in a vicarage where he had gone seeking a handout. He was convicted of murder and sentenced to hang, but his sentence was commuted to life imprisonment.

Another case is that of Raymond Tanner, a convicted sex offender, who pleaded guilty to the beating and rape of a woman in California. He is 6 feet 3 inches tall, mentally disordered, and has an XYY complement. A superior court judge is attempting to decide whether Tanner's plea of guilty to assault with intent to commit rape will stand, or whether he will be allowed to plead innocent by reason of insanity.

Criminal lawyers in the United States have already begun to request genetic studies of their clients. In October of 1968 a lawyer for Sean Farley, a 26-year-old XYY man in New York who was charged with a rape-slaying, maneuvered to raise the issue of his client's genetic defect in court.

Many questions are raised by the double Y syndrome—basic social, legal and ethical questions—which will become more and more insistent as the implications of chromosome abnormalities take root in the public mind. Is an extra Y chromosome causally related to antisocial behavior? Is there a genetic basis for criminal behavior? If a man has an inborn tendency toward criminal behavior, can we fairly hold him legally accountable for his acts? If a criminal's chromosomes are at fault, how can we rehabilitate him?

The evidence to date is inadequate to prove conclusively the validity of the syndrome and convict all of the world's estimated five million XYY males of innate aggressive or criminal tendencies. But if the concept is proved, what then? The first step would seem to be to identify the XYY infants in the general population. This suggests the need for a nationwide program of automatic chromosome analysis of all newborns.

How should society deal with XYY individuals? If they are genetically abnormal, they should not be treated as normal. If the XYY condition dooms a man to a life of crime, he should be restrained but not punished. Mongolism also is a chromosome abnormality, and afflicted individuals are not held responsible for their behavior. Some valuable suggestions on the legal aspects of the double Y syndrome have been published recently by

Kennedy McWhirter[59]. Elsewhere, Kessler and Moos[51] claim that definitive concepts relating to the YY syndrome have been accepted prematurely.

If all infants could be karyotyped at birth or soon after, society could be forearmed with information on chromosome abnormalities and perhaps it could institute the proper preventive and other measures at an early age. Although society can not control the chromosomes (at least at the present time) it can do a great deal to change certain environmental conditions that may encourage XYY individuals to commit criminal acts.

The theory that a genetic abnormality may predispose a man to antisocial behavior, including crimes of violence, is deceptively and attractively simple, but will be difficult to prove. Extensive chromosome screening with prospective follow-up of XYY males will be essential to determine the precise behavioral risk of this group. It is by no means universally accepted yet. Many geneticists urge that we should be cautious in accepting the interpretation that the double Y condition is specifically associated with criminal behavior, and particularly so with reference to the medicolegal validity of these concepts.

Literature Cited

1   Anonymous, The YY syndrome. *Lancet* 1:583-584. 1966.

2. —————. Criminal behavior—XYY criterion doubtful. *Science News* 96:2. 1969.

3. Baikie, A. G., W. M. Court Brown, K. E. Buckton, D. G. Harnden, P. A. Jacobs and I. M. Tough. A possible specific chromosome abnormality in human chronic myeloid leukemia. *Nature* 188:1165-1166. 1960.

4. —————, O. Margaret Garson, Sandra M. Weste, and Jean Ferguson. Numerical abnormalities of the X chromosome. *Lancet* 1:398-400. 1966.

5. Balodimos, Marios C., Hermann Lisco, Irene Irwin, Wilma Merrill, and Joseph F. Dingman. XYY karyotype in a case of familial hypogonadism. *J. Clin. Endocr.* 26:443-452. 1966.

6. Barr, M. L. Sex chromatin and phenotype in man. *Science* 130:679. 1959.

7. —————— and D. H. Carr. Sex chromatin, sex chromosomes and sex anomalies. *Canad. Med. Assn. J.* 83:979-986. 1960.

8. ——————, D. H. Carr, H. C. Soltan, Ruth G. Wiens, and E. R. Plunkett. The XXYY variant of Klinefelter's syndrome. *Canad. Med. Assn. J.* 90:575-580. 1964.

9. Belsky, Joseph L. And George H. Mickey. Human cytogenetic studies. *Danbury Hospital Bull.* 1:19-20. 1965.

10. Boczkowski, K. and M. D. Casey. Pattern of DNA replication of the sex chromosomes in three males, two with XYY and one with XXYY karyotype. *Nature* 213:928-930. 1967.

11. Buckton, Karin E., Jane A. Bond, and J. A. McBrid e. An XYY sex chromosome complement in a male with hypogonadism. *Human Chromosome Newsletter* No. 8, p. 11. Dec. 1962.

12. Carakushansky, Gerson, Richard L. Neu, and Lytt I. Gardner. XYY with abnormal genitalia. *Lancet* 2:1144. 1968.
13. Carr, D. H. Chromosome studies in abortuses and stillborn infants. *Lancet* 2:603-606. 1963.
14. _____, M. L. Barr, and E. R. Plunkett. An XXXX sex chromosome complex in two mentally defective females. *Canad. Med., Assn. J.* 84-131-137. 1961.
15. Casey, M. D., C. E. Blank, D. R. K. Street, L. J. Segall, J. H. McDougall, P. J. McGrath, and J. L. Skinner. YY chromosomes and antisocial behavior. *Lancet* 2:859-860. 1966.
16. _____, L. J. Segall, D. R. K. Street, and C. E. Blank. Sex chromosome abnormalities in two state hospitals for patients requiring special security. *Nature* 209:641-642. 1966.
17. _____, D. R. K. Street, L. J. Segall, and C. E. Blank. Patients with sex chromatin abnormality in two state hospitals. *Ann. Human Genet.* 32:53-63. 1968.
18. Close, H. G., A. S. R. Goonetilléke, Patricia A. Jacobs, and W. H. Price. The incidence of sex chromosomal abnormalities in mentally subnormal males. *Cytogenetics* 7:277-285. 1968.
19. Conan, P. E. and Bayzar Erkman. Frequency and occurrence of chromosomal syndromes. I. D-trisomy. *Am. J. Human Genet.* 18:374-386. 1966.
20. _____ and _____. Frequency and occurrence of chromosomal syndromes. II. E-trisomy. *Am. J. Human Genet.* 18:387-398. 1966.
21. Court Brown, W. M. Sex chromosomes and the law. *Lancet* 2:508-509. 1962.
21a._____. Males with an XYY sex chromosome complement. *J. Med. Genet.* 5:341-359. 1968.
22. Cowie, John and Jacob Kahn. XYY constitution in prepubertal child. *Brit. Med. J.* 1:748-749. 1968.
23. deCapoa, A., D. Warburton, W. R. Breg, D. A. Miller, and O. J. Miller. Translocation heterozygosis: a cause of five cases of *cri du chat* syndrome and two cases with a duplication of chromosome number five in three families. *Am. J. Human Genet.* 19:586-603. 1967.
24. Dent, T., J. H. Edwards, and J. D. A. Delhanty. A partial mongol. *Lancet* 2:484-487. 1963.
25. Edwards, J. H., D. G. Harnden, A. H. Cameron, V. Mary Crosse, and O. H. Wolff. A new trisomic syndrome. *Lancet* 1:787-790. 1960.
26. Eggen, Robert R. Chromosome Diagnostics in Clinical Medicine. Charles C. Thomas, Springfield, Ill. 1965.
27. Ferguson-Smith, M. A., Marie E. Ferguson-Smith, Patricia M. Ellis, and Marion Dickson. The sites and relative frequencies of secondary constrictions in human somatic chromosomes. *Cytogenetics* 1:325-343. 1962.
28. Ford, C. E. and J. L. Hamerton. The chromosomes of man. *Nature* 178:1020-1023. 1956.
29. _____, K. W. Jones, O. J. Miller, Ursula Mittwoch, L. S. Penrose, M. Ridler, and A. Shapiro. The chromosomes in a patient showing both mongolism and the Klinefelter syndrome. *Lancet* 1:709-710. 1959.
30. _____, K. W. Jones, P. E. Polani, J. C. deAlmedia, and J. H. Briggs. A sex-chromosome anomaly in a case of gonadal dysgenesis (Turner's syndrome). *Lancet* 1:711-713. 1959.

31. _____, P. E. Polani, J. H. Briggs, and P. M. F. Bishop. A presumptive human XXY/XX mosaic. *Nature* 183:1030-1032. 1959.

32. Forssman, H. and G. Hambert. Incidence of Klinefelter's syndrome among mental patients. *Lancet* 1:1327. 1963.

33. Fraccaro, M., K. Kaijser, and J. Lindsten. Chromosome complement in gonadal dysgenesis (Turner's syndrome). *Lancet* 1:886. 1959.

34. _____, M. Glen Bott, P. Davies, and W. Schutt. Mental deficiency and undescended testia in two males with XYY sex chromosomes. *Folia Hered. Pathol. (Milan)* 11:211-220. 1962.

35. Franks, Robert C., Kenneth W. Bunting, and Eric Engel. Male pseudohermaphrodism and XYY sex chromosomes. *J. Clin. Endocr.* 27:1623-1627. 1967.

36. Fraser, J. H., J. Campbell, R. C. MacGillivray, E. Boyd, and B. Lennox. The XXX syndrome—frequency among mental defectives and fertility. *Lancet* 2:626-627. 1960.

37. Gates, William H. A case of non-disjunction in the mouse. *Genetics* 12:295-306. 1927.

38. Hamerton, J. L. Sex chromatin and human chromosomes. *Intern. Rev. Cytol.* 12:1-68. 1961.

39. Hauschka, Theodore S., John E. Hasson, Milton N. Goldstein, George F. Koepf, and Avery A. Sandberg. An XYY man with progeny indicating familial tendency to non-disjunction. *Am. J. Human Genet.* 14:22-30. 1962.

40. Hayward, M. D. and B. D. Bower. Chromosomal trisomy associated with the Sturge-Weber syndrome. *Lancet* 2:844-846. 1960.

41. Hunter, H. Chromatin-positive and XYY boys in approved schools. *Lancet* 1:816. 1968.

42. Hustinx, T. W. J. and A. H. F. van Olphen. An XYY chromosome pattern in a boy with Marfan's syndrome. *Genetica* 34:262. 1963.

43. Ismail, A. A. A., R. A. Harkness, K. E. Kirkham, J. A. Loraine, P. B. Whatmore, and R. P. Brittain. Effect of abnormal sex-chromosome complements on urinary testosterone levels. *Lancet* 1:220-222. 1968.

44. Jacobs, P. A., A. G. Baikie, W. M. Court Brown, and J. A. Strong. The somatic chromosomes in mongolism. *Lancet* 1:710. 1959.

45. _____, and A. J. Kay. Chromosomes in a child with Bonnevie-Ullrich syndrome. *Lancet* 2:732. 1959.

46. _____, and J.A. Strong. A case of human intersexuality having a possible XXY sex-determining mechanism. *Nature* 182: 302-303. 1959.

47. _____, A. G. Baikie, W. M. Court Brown, T. N. MacGregor, N. MacLean, and D. G. Harnden. Evidence for the existence of the human "super female". *Lancet* 2:423-425. 1959.

48. _____, Muriel Brunton, Marie M. Melville, R. P. Brittain, and W. F. McClemont. Aggressive behavior, mental subnormality and the XYY male. *Nature* 208:1351-1352. 1965.

49. _____, W. H. Price, W. M. Court Brown, R. P. Brittain, and P. B. Whatmore. Chromosome studies on men in a maximum security hospital. *Ann. Human Genet.* 31:330-347. 1968.

50. Kesaree, Nirmala and Paul V. Woolley. A phenotypic female with 49 chromosomes, presumably XXXXX. *J. Pediat.* 63:1099-1103. 1963.

51. Kessler, Seymour and Rudolph H. Moos. XYY chromosome: premature conclusions. *Science* 165:442. 1969.

52. Kosenow, W. and R. A. Pfeiffer. YY syndrome with multiple malformations. *Lancet* 1:1375-1376. 1966.

53. Leff, J. P. and P. D. Scott. XYY and intelligence. *Lancet* 1:645. 1968.

54. Lejeune, J., M. Gautier, and R. Turpin. Etude des chromosomes somatiques de neuf enfants mongoliens. *Compt. Rend. Acad. Sci.* 248:1721-1722. 1959.

55. _____, J. Lafourcade, R. Berger and M. O. Rethore. Maladie du cri du chat et sa reciproque. *Ann. Genet.* 8:11-15. 1965.

56. Lubs, H. A., Jr., E. V. Koenig, and L. H. Brandt. Trisomy 13-15: A clinical syndrome. *Lancet* 2:1001-1002. 1961.

57. Lyon, M. F. Gene action in the X-chromosome of the Mouse (*Mus musculus* L.*).* *Nature* 190:372-373. 1961.

58. Maclean, N., D. G. Harnden, W. M. Court Brown, Jane Bond, and D. J. Mantle. Sex-chromosome abnormalities in newborn babies. *Lancet* 1:286-290. 1964.

59. McWhirter, Kennedy. XYY chromosome and criminal acts. *Science* 164:1117. 1969.

59a. Melnyk, John, Frank Vanasek, Havelock Thompson, and Alfred J. Rucci. Failure of transmission of supernumerary Y chromosomes in man. Abst. Am. Soc. Human Genet. Annual Meeting, Oct. 1-4, 1969.

60. Mickey, George H. Chromosome studies in testicular feminization syndrome in human male pseudohermaphrodites. *Mammalian Chromosome Newsletter* No. 9, p. 60. 1963.

61. Migeon, Barbara R. G trisomy in an XYY male. *Human Chromosome Newsletter* No. 17. Dec. 1965.

62. Milcu, M., I. Nigoescu, C. Maximilian, M. Garoiu, M. Augustin, and Ileana Iliescu. Baiat cu hipospadias si cariotip XYY. *Studio si Cercetari de Endocrinologie* (Bucharest) 15:347-349. 1964.

63. Miller, Orlando J. The sex chromosome anomalies. *Am. J. Obstet. Gynec.* 90:1078-1139. 1964.

64. Minckler, Leon S. Chromosomes of criminals. *Science* 163;1145. 1969.

65. Montagu, Ashley. Chromosomes and crime. *Psychology Today* 2:43-49. 1968.

66. Muldal, S. and C. H. Ockey. The "double male": a new chromosome constitution in Klinefelter's syndrome. *Lancet* 2:492-493. 1960.

67. Nowell, P. C. and D. A. Hungerford. A minute chromosome in human granulocytic leukemia. *Science* 132: 1497. 1960.

68. Painter, T. S. The chromosome constitution of Gates "non-disjunction" (v-o) mice. *Genetics* 12:379-392. 1927.

68a. Palmer, Catherine G. and Sandra Funderburk. Secondary constrictions in human chromosomes. *Cytogenetics* 4:261-276. 1965.

69. Patau, K. The identification of individual chromosomes, especially in man. *Am. J. Human Genet.* 12:250-276. 1960.

70. _____, D. W. Smith, E. Therman, S. L. Inhorn, and H. P. Wagner. Multiple congenital anomalies caused by an extra chromosome. *Lancet* 1:790-793. 1960.

71. Penrose, L. S. The Biology of Mental Defect. Grune and Stratton, New York. 1949.

72. Pergament, Eugene, Hideo Sato, Stanley Berlow, and Richard Mintzer. YY

syndrome in an American negro. *Lancet* 2:281. 1968.

73. Pfeiffer, R. A. Der Phanotyp der Chromosomenaberration XYY. *Wochenschrift* 91:1355-1256. 1966.

74. Price, W. H. and P. B. Whatmore. Behaviour disorders and the pattern of crime among XYY males identified at a maximum security hospital. *Brit. Med. J.* 1:533. 1967.

75. _____, J. A. Strong, P. B. Whatmore, and W. F. McClement. Criminal patients with XYY sex-chromosome complement. *Lancet* 1:565-566. 1966.

76. Puck, T. T., A. Robinson, and J. H. Tjio. A familial primary amenorrhea due to testicular feminization. A human gene affecting sex differentiation. *Proc. Exper. Biol. Med.* 103:192-196. 1960.

77. Reitalu, Juhan. Chromosome studies in connection with sex chromosomal deviations in man. *Hereditas* 59:1-48. 1968.

78. Ricci, N. and P. Malacarne. An XYY human male. *Lancet* 1:721. 1964.

79. Sandberg, A. A., G. F. Koepf, T. Ishihara, and T. S. Hauschka. An XYY human male. *Lancet* 2:488-489. 1961.

80. _____, Takaaki Ishihara, Lois H. Crosswhite, and George F. Koepf. XYY genotype. *New England J. Med.* 268:585-589. 1963.

81. Sergovich, F., G. H. Valentine, A. T. L. Chem, R. A. H. Kinch, and M. S. Smout. Chromosome aberrations in 2159 consecutive newborn babies. *New England J. Med.* 280:851-855. 1969.

82. Smith, D. W., K. Patau, and E. Therman. The 18 trisomy syndrome and the D_1 trisomy syndrome. *Am. J. Dis. Child.* 102:587. 1961.

83. _____, K. Patau, E. Therman, S. L. Inhorn, and R. I. Demars. The D_1 trisomy syndrome. *J. Pediat.* 62:326-341. 1963.

84. Sohval, Arthur R. Sex chromatin, chromosomes and male infertility. *Fertility and Sterility* 14:180-207. 1963.

85. _____. Chromosomes and sex chromatin in normal and anomalous sexual development. *Physiol. Rev.* 43:306-356. 1963.

86. Telfer, Mary A., David Baker, Gerald R. Clark, and Claude E. Richardson. Incidence of gross chromosomal errors among tall criminal American males. *Science* 159:1249-1250. 1968.

87. Thorburn, Marigold J., Winston Chutkan, Rolf Richards and Ruth Bell. XYY sex chromosomes in a Jamaican with orthopaedic abnormalities. *J. Med. Genet.* 5:215-219. 1968.

88. Tjio, J. H. and A. Levan. The chromosome number of man. *Hereditas* 42:1-6. 1956.

89. _____, T. T. Puck, and A. Robinson. The somatic chromosomal constitution of some human subjects with genetic defects. *Proc. Natl. Acad. Sci.* 45:1008-1016. 1959.

90. Townes Philip L., Nancy A. Ziegler, and Linda W. Lenhard. A patient with 48 chromosomes (XYYY). *Lancet* 1:1041-1043. 1965.

91. Vignetti, P., L. Capotorti, and E. Ferrante. XYY chromosomal constitution with genital abnormality. *Lancet* 2:588-589. 1964.

92. Welch, J. P., D. S. Borgaonkar, and H. M. Herr. Psychopathy, mental deficiency, aggressiveness and the XYY syndrome. *Nature* 214:500-501. 1967. 93.

93. Wiener, Saul, and Grant Sutherland. A normal XYY man. *Lancet* 2:1352. 1968

4
Getting Down to Business in Genetics Research, Diagnosis, Treatment

PAUL T. LIBASSI

There are good reasons why there has been such a jump in molecular biology as applied to bacterial viruses and bacteria. One of these reasons, and possibly the most important, is the fact that these are problems on which work can be done, by and large, by one man in a small laboratory with limited equipment. The nature of the universities and the nature of most research institutions are of such quality that they insist a man work precisely in this way. Therefore, large problems, large problems that involve coordinated efforts, very often are not pursued because the facilities and funding and cooperation are not organized within the universities and research institutions; the result is that certain kinds of large problems are selected against.

Thus, Dr. Seymour Cohen of the University of Pennsylvania summarized the cardinal reason behind genetic science's failure to advance substantially beyond the plateau of bacteria and bacteriophages. Financial requisites and organizational requirements must be established, he emphasized at a National Institute of General Medical Sciences symposium over two years ago, if the thrust of genetics research is to progress beyond the realm of lower organisms toward more pertinent vistas.

That inherited diseases of man are indeed such a "large problem" is attested to by medical statistician Gabriel Stickle, who has calculated the numbers of life years lost due to birth defects—both inborn errors of

Reprinted from *Laboratory Management*. Copyright United Business Publication Inc. 1972.

metabolism and structural deformities. Stickle has projected over 36 million life years lost in 1967 alone. According to that statistic, birth defects claim approximately 4.5 times as many life years as heart disease, 8 times as many as cancer, and about 10 times as many as stroke.

Facilities and cooperation, financial requisites and organizational requirements, remarked Dr. Cohen, are required to move the course of genetics research forward. Through grants totaling in excess of $3.3 million, NIGMS this past June established and funded a network of seven Genetics Centers in five states (see box below). Interdisciplinary analysis, cooperative endeavors, aggregates of highly coordinated and integrated research projects characterized the strategem of this research conglomerate. The design of each cetner prescribes that teams of scientists and medical geneticists explore the basic molecular nature of genetic diseases concurrent with clinical evaluations and population analyses involving affected patients and members of their families.

Studies at these Centers will be augmented by research findings generated by some 300 smaller projects which NIGMS also supports, bringing its total program expenditure to approximately $30 million per year. From these smaller projects—the bulk of which deal with genetic investigation of lower organisms—come the concepts and prototypes for the exploration of human genetics.

Centers vs. Grants

Why does a funding instrument make the decision to found a Center requiring hundreds of thousands of dollars to support, when a smaller project can be sustained through an endowment of some $10,000 or $20,000? In a series of interviews with Dr. Fred H. Bergmann, Chief of the Genetics Section, Research Grants Branch at NIGMS, the answers to that and other questions regarding the evolution and rationale of these Centers were discussed.

The reasons why these Centers were established can be summarized under two general headings: *managerial* and *scientific*.

"The scientific potentialities and the talent to exploit them," explains Dr. Bergmann, "existed before NIGMS decided to fund these Centers. The problem was to catalyze these talents to interact as a group, to convince each separate discipline to function together. NIGMS is the glue that cements these parts into a whole."

But problems of management extend well beyond the absence of interdisciplinary communications and interaction. Difficulties that stem from the nature of individualized research—Dr. Cohen calls them sociological problems in the discovery-making process—must be contended with. The collaborative design of these seven Genetics Centers may well have done just that.

When dealing, for instance, with lower organisms such as *E. Coli,* which the majority of genetics research to date has, work can be done and great progress made by "one man in a small laboratory with limited equipment." The structural intricacies and complexities of human genetics, however, defy research of this type. They demand the professional acumen of an assemblage of specialists from the molecular biologist to the biochemist, enhanced by a sophisticated battery of equipment and facilities—equipment and facilities which can easily exceed economic feasibility under a $10,000 stipend.

NIH Genetics Centers Program

Center	Principal Investigator	Disease Foci
Albert Einstein College of Medicine, Yeshiva University, New York ($319,300)	Harold M. Nitowsky, M.D.	Blood disorders; hereditary factors in diabetes
The Johns Hopkins University, Baltimore ($570,455)	Victor McKusick, M.D.	Bone and skeletal disorders; Marfan syndrome (heart enlargement, aortic weakness)
Mount Sinai School of Medicine, New York ($331,425)	Kurt Hirschhorn, M.D.	Hereditary kidney diseases; genetic eye diseases in children
University of California School of Medicine, San Diego ($719,659)	J. Edwin Seegmiller, M.D.	Metabolic disorders (Lesch-Nyhan, Tay-Sachs diseases); polysaccharide disorders (Hurler's syndrome)
University of California Medical Center at San Francisco ($480,793)	Charles J. Epstein, M.D.	Sex chromosome defects (Turner's syndrome); gout (disease of purine metabolism)
University of Texas Graduate School of Biomedical Sciences, Houston ($583,047)	Margery W. Shaw, M.D.	Carbohydrate storage disease; cystic fibrosis (identification of carriers)
University of Washington School of Medicine, Seattle ($958,207)	Arno G. Motulsky, M.D.	Sickle cell anemia (cyanate therapy); hereditary lipid patterns in heart diseases

Table courtesy of the National Institutes of Health

"You simply cannot justify expensive computer time or a tissue culture laboratory when you are dealing with a $10,000 grant," says Dr. Bergmann. To justify assests and facilities that run into large sums of money and the finances to implement them, he continues, interdisciplinary efforts concerned with broader issues and involving groups of scientists must be established first.

Finances of the magnitude necessary to sustain collaborative research on such a large scale do, however, demand more meticulous surveillance and management than do lesser individual endowments. Who accounts for expenditures? Who is held responsible in the event that affairs run amuck? In short, who is the watchdog?

The onus of administrative accountability falls upon the NIGMS-designated "principal investigator" of each Center. Correspondingly, each "principal" is free to delegate whatever administrative responsibility he or she feels is most advantageous to his or her specific program. The mood and tone of each Center is set by the "principal;" from that point, it evolves into a cooperative venture.

Scientific rationale.

In the last several years, genetics research has stood at the brink of change. The magnificent portfolio of genetic information gleaned from lower organisms has set the stage for manipulative genetics. It is, however, virtually impossible to develop technologies for manipulative genetics in humans outside the scope of interdisciplinary research. What is required, what these Centers provide, are cores of geneticists surrounded by biochemists, molecular biologists, physicists and a myriad of other specialists.

The scientific rationale behind these Centers follows a logical progression from the simple to the complex, from what is already known to precisely what must be found out. Just as information derived from the scrutiny of lower organisms affords points of departure for research into higher organisms, these Centers propose to develop the most rudimentary genetic disorders as prototypes for the understanding and treatment of more complex genetic defects. This philosophy makes the ultimate goal of each Center a comprehension of polygenic defects, while the focus, at least for the moment, is on understanding the apparatus of Mendelian diseases.

Genetically related diseases may be categorized in order of ascending complexity according to the following scheme. The first and most elementary type of defect occurs when a single Mendelian gene is mutant and produces a disease such as hemophilia. The machinery responsible for the majority of Mendelian defects is a chemical alteration in the DNA causing an amino acid substitution in an end-product protein.

Chromosomal aberrations or abnormalities—Turner's syndrome, for example—represent the second category of disorders. "Here, generally," explains Dr. Arno Motulsky, principal investigator at the University of Washington Center, "the abnormality is not transmitted, but occurs as a *de nova* event in the sperm cells of the parents or in the first few divisions of the fertilized egg." As in the case of mongolism and other types of mental retardation, deleterious genetic material latent in the parents surfaces with profound consequences in the offspring.

The mechanisms of Mendelian defects and the pathways through which chromosomal abnormalities produce multiple clinical aberrations spearhead research at these seven Centers. Much is presently known about the machinery of single-gene diseases. The study of chromosomal pathways is, in contrast, still in germinal stages. And though the incidence and social consequences of these so-called Mendelian diseases seem monumental, they appear to shrink against the backdrop of polygenic disorders.

The third or polygenic category of genetic defects is that in which multiple genes interact to somehow cause an individual to be more susceptible to a given disease. Certain genes have, in fact, been implicated. in a susceptibility to certain diseases. An entire spectrum of disorders ranging from atherosclerosis to heart disease to diabetes and schizophrenia probably have such a mechanism operative, according to Dr. Motulsky. "Case findings in patients and particularly in their families, therefore, have great public health significance."

Where funds are limited and resources finite, a multi-million dollar research network, according to Dr. Bergmann, cannot be justified on the basis of diseases that are comparatively "minor" in terms of their social impact. What NIGMS appears to be banking on is that an understanding of the lower members of the genetic disease hierarchy will yield models for the treatment and cure of polygenic disorders.

Therapeutic Hierarchy

The shared and ultimate goal of each of these Centers is the development of technologies by which the basic hereditary matarial DNA may be harnessed as therapeutic agent for alleviating genetic diseases.

"If we can deliver the DNA that could cure or treat some of these disorders," explains Dr. Vasken Aposhian of the University of Maryland School of Medicine, "then the mechanism of delivering the DNA could be useful in a large number of diseases over the entire span of life-time of various individuals." That same mechanism of delivery—"not what is inside, but the mechanism of delivery," stresses Dr. Aposhian—may well be applicable to the entire gamut of genetic disorders.

A broad spectrum of inborn metabolic errors eventuate when faulty information in DNA orders up inadequate amounts of a particular enzyme. Corrective, or at least therapeutic, strategy for these anomalies proceeds from the enzymatic reaction: substrate is converted to a product and is catalyzed by an enzyme.

According to this reaction, treatment can be one or more of the following procedures: (1) remove the accumulated substrate; (2) add product of the blocked reaction; (3) add enzyme; (4) add the missing gene(s) which direct synthesis of enzyme.

Therapies derived from this equation are in some cases now available—specifically in adding products or removing substrates. In other cases they remain largely a matter of optimistic speculation.

Measures one and two are, by and large, according to Dr. Bergmann, well documented. The treatment for hypopituitary dwarfism with growth hormone, for instance, is an example of the addition of a product in a blocked reaction. Another example is the use of insulin for the treatment of juvenile diabetes. In patients suffering from phenylketonuria, a prodigous volume of phenylalanine and phenylpyruvic acid accumulates. In such amounts, both of these compounds are, for reasons unknown, toxic. The condition can be alleviated through control of dietary intake thereby eliminating the accumulated substrate or cofactors.

"In a few instances," points out Dr. DeWitt Stetten, Jr., Director of NIGMS, "some of the symptoms of the diseases can be deferred or prevented by drastic dietary measures or by replacement of the end-product protein. But, none of these diseases can yet be cured in the sense that a victim no longer needs continuing treatment."

What is needed, what these Centers hope to construct, is a bridge between molecular biology and human biology. The addition of an enzyme to a system, or in other cases, of the missing gene or genes which direct synthesis of the enzyme that is deficient are the kingpins of such a bridge.

The use of enzymes in therapeutics has been comparatively unsuccessful in mammalian systems over the years, according to Dr. Aposhian. Several experiments with laboratory animals do, however, provide cases where enzyme catalase encase in plastic microcapsules permitted enzymes to diffuse gradually into cellular mediums with encouraging results. "Whether this will become a uniform method of treatment," explains Dr. Aposhian, "is of question because of the kinds of enzymes involved and the size of enzymes. It certainly is worthy of investigation." Researchers in at least two of the Genetics Centers—the University of California Medical Center at San Francisco and the University of Texas Graduate School of Biomedical Sciences—plan to do just that.

Transduction and Transformation

The fourth, and most speculative and controversial method of treating inborn errors of metabolism—addition of the missing gene (s) or "genetic engineering"—may also prove itself to be the most promising. There is a need for DNA as a therapeutic agent because, says Dr. Aposhian, some of these diseases cannot be treated in any other way successfully.

How can a gene be introduced into a mammalian cell that is deficient in that gene? And assuming that the gene can be implanted, and that its genetic code corresponds to that which is deficient in the host, how can one guarantee that it will arrive at its intended destination, the liver for example, intact?

Explains Dr. Aposhian, "One thing we know from molecular biology is that we can give mammalian cell a piece of DNA and have that DNA expressed genetically in its own fashion if we protect that DNA by means of a protein coat. This is essentially what a virus is, a nucleic acid surrounded and protected by a protein coat. We do have model systems of transformation of bacterial systems, of infecting the bacterial cells with naked DNA or DNA surrounded by something."

Prospects for a potential "mechanism of delivery" stem from a phenomenon in molecular biology called transduction—the delivery of a piece of host DNA inside a virus to a cellular recipient. Model systems available from successful implants in microbial genetics suggest lanes of approach which may be applicable to mammalian virology. Cause for optimism originates from experiments which demonstrate that when mouse kidney cells are infected with polyoma virus, the resultant progeny display evidence of total host DNA genome, as well as polyoma DNA. Says Dr. Aposhian, "This was the first time that this type of phenomena has been seen in animal virology. We think we now may have something that will help us here."

The log of questions yet to be answered looms before these Centers in well-defined but Herculean perspective. Easy and reproducible methods for gene isolation, synthetic techniques for DNA manufacture and production, and procedures to segregate and identify mammalian genes are among the most pressing issues. Until further studies substantiate that pseudovirus DNA is actually being physically integrated into host chromosomes (and not being "chopped up" by enzymes in the cell), pseudovirus must be administered at regular intervals, about every two to four weeks. But what of antibody synthesis, the antigenicity of the viral particle?

"What you eventually need for human gene therapy, speculated Dr. Aposhian in 1970, "if you jump now ten or fifteen years or whatever it is, is human viruses that have low antigenicity and are nonpathogenic."

Patients and Physicians

To what degree these seven Centers will entertain diagnostic and consultative dialogue with patients and physicians is only tacitly implied in the arithmetic of these grants. Dr. Bergmann, when questioned regarding guidelines for inpatient selection and treatment, explained it thus: The mandate of these Centers is not to provide health services, but to stimulate an environment for collaborative research. It all boils down to a question of emphasis. Scientists working at these Centers are paid to do research. There will, of course, be a coincident interaction between scientists and physicians. Diagnostic consultation and patient counseling and treatment will result, and that is all well and good. The point remains, however, that these Centers were founded and exist as research facilities. Counseling and education are spin-off products and are not necessarily basic to the design of each Center. Specific Centers may elect to include these as components of their programs. How integral a role counseling and education will play is up to these institutions.

There are instances where genetics research cannot proceed exclusive of a patient population. Recognition and diagnosis of carrier states, for example, involve both patients and physicians through exhaustively detailed case histories and on-site examinations.

"We are very interested," explains Dr. Margery Shaw, principal investigator of the Medical Genetics Center (MGC) at the University of Texas, "in studying individuals who carry the gene but who don't manifest the disease," To this end, "any" patient with genetic disease, and relatives, will be of paramount concern to the MGC clinical unit.

Because NIGMS intends its allocations to fund research, only small monetary provisions have been earmarked to defray hospitalization expenses incurred by patients. Dr. Bergmann cautioned that such a measure is not meant to discourage health services, but to underscore a research priority. NIGMS foresees most expenses being subsidized through third party funds, such as Blue Cross, Blue Shield, and other hospitalization insurance.

Genetic Cartographers

The microworlds and mini-frontiers that genetic scientists will pioneer at these seven Centers pose vast uncharted terrains. To discuss the scope and magnitude of the genetic enigmas that each will explore is, according to Dr. Bergmann, almost inadequate outside the context of a 300-page thesis. What follows are profiles of experimental activity at these Centers. These sketches are presented as being representative of the type and nature of research which will take place. They are not, in any case, meant to cover the entire range of projects.

Chromosomal pathways remain a mystery that genetic science has only recently begun to solve. At Johns Hopkins University, scientists will "map" the expression of mammalian genes. Investigators hope to identify and align genes on a linear map according to their individual characteristics and the frequency with which these characteristics are combined and expressed in offspring. As an adjunct to this study, researchers plan a study in population dynamics. because genetic disorders manifest themselves more readily in inbred groups, genetic statisticians will gather data from sects of Amish people, who according to custom and religious conviction rarely wed outside the boundaries of their immediate communities.

"Up until about one year ago," Dr. Bergmann explains, "chromosomes could be distinguished only by shape. A chromosome was fat, thin, in the shape of a cross, whatever." Today, with the advent of more sophisticated staining and fluorescent procedures, scientists at the University of Texas Center plan to specify chromosomes according to banding techniques—banding patterns are unique for each particular human chromosome. "I am very interested in the possibility that chromosomes have inherent banding patterns in different individuals," says Dr. Shaw.

Dr. William J. Schull, an associate of Dr. Shaw will pursue demographic studies to ascertain the prevalence of genetic disease among Negroes,

Whites and Chicanos in the Houston area. A corollary project proposes to determine any correlation which may exist between a propensity towards genetic disease and certain situations and environments. Through computer simulation, Dr. Schull will trace through several generations precisely what effects might result from various modes of family planning.

Three departments—Pediatrics, Medicine and Biochemistry—at the University of California Medical Center at San Francisco will be directly involved in the Genetics Center there. Investigations will deal primarily with general and clinical problems related to chromosomal abnormalities and metabolic defects. Research will also be conducted on such basic problems as the mechanics of gene action, genetic control of metabolism and development, and the effects of genetic abnormalities on these functions. An important component of research there, according to Dr. Charles J. Epstein, principal investigator, will be the establishment of a centralized laboratory for growing cells in tissue culture, an asset important to both basic and clinical studies to be undertaken.

At the University of Washington School of Medicine in Seattle, scientists will investigate such diverse problems as the possible use of cyanate therapy for sickle cell anemia and the genetics of hyperkinetic children. Researchers also hope to develop more efficient technologies for pre-natal diagnosis and mass screening for inborn errors of metabolism.

Treatment of storage diseases through plasma transfusion, metabolic defects in red blood cells and hemoglobin variations, and hereditary factors in diabetes comprise the core of research problems to be studied at the Albert Einstein Genetics Center in New York City. Scientists there are also interested in mapping genes and chromosomes in an effort to determine data on translocation—what enzymes are involved with what chromosomes. Dr. Bergmann terms this latter project "very basic to an understanding of genetics."

Myopia or near-sightedness intrigues researchers at the Mount Sinai School of Medicine in New York because of its primary incidence among children of Puerto Rican descent. "Studies of this sort," explains Dr. Bergmann, "point up the interdisciplinary nature of these Centers very well. To research myopia, for example, will require close collaboration with an optometrist."

III
TECHNIQUES OF GENETIC ENGINEERING

Detecting and Correcting Genetic Defects vs. Modifying the Human Gene Pool

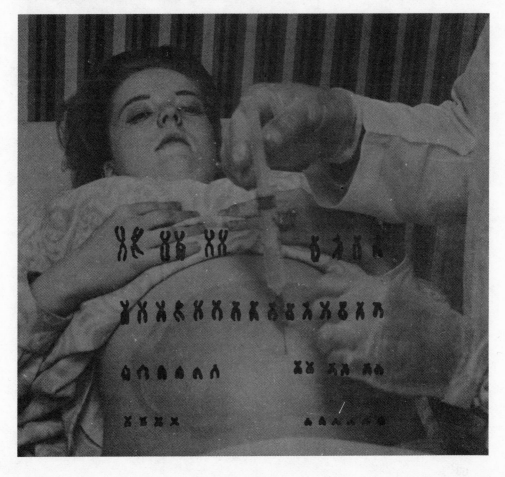

It should be perfectly clear to anyone who gives the issue even casual consideration that human beings are intervening more and more in their own natural selection process. For example, twentieth century *Homo sapiens* clearly is not subjected to the same selection pressures that were virtually certain to eliminate particular genotypes in the nineteenth century. Today, the diabetic can survive with supplemental insulin; the hemophiliac can be supplied with the blood component that his genotype will not permit his own body to produce; and the newborn with phenylketonuria, who is placed on a diet lacking in the offending amino acid phenylalanine, will develop without the debilitating mental retardation typical of PKU. All of us willingly accept advances in medical science that enable genetically defective individuals to live more normal and productive lives. At the same time it should be emphasized that the long-term effect of such medical advances must be the accumulation in the human gene pool of increasing numbers of deleterious genes.

Let there be no mistake about it, then, a kind of "genetic engineering" is one of the inherent consequences of many advances in medical science. Similarly, genetic counseling that results in a couple's decision not to have children when both are heterozygous for the gene for Tay-Sachs' disease or for the gene for sickle cell anemia, may also affect the human gene pool. In contrast to these types of "genetic engineering," which are concerned with the welfare of individuals and families, are those practices or potential practices that are designed specifically to modify the human gene pool and thus to regulate the evolutionary future of mankind. Generally little controversy surrounds a procedure designed to benefit an individual patient and his or her family, even when the procedure may have long-term detrimental effects on the human gene pool. By way of contrast, a storm of protest often is aroused by advocacy of a "genetic engineering" proposal that is not concerned with the individual patient, but that is designed to improve the human gene pool.

In this section of the book you will be reading about both kinds of genetic and reproductive engineering procedures. Some of the procedures, such as genetic counseling and amniocentesis are currently being practiced and are of significance to the alleviation of suffering by genetically defective individuals or their families. Other procedures (e.g., transplantation of normal liver tissue into a genetically defective individual who cannot produce a particular liver enzyme) are being tested with animals and have the potential for alleviating suffering in genetically defective humans. However, still other procedures, such as Muller's "voluntary choice of germ plasm" and cloning are advocated by those whose primary goal is engineering improvements in the human gene pool and thereby enhancing the evolutionary future of *Homo sapiens.*

This section of the book includes 11 diverse articles concerned with currently or potentially feasible ways in which gene defects may be corrected in individuals and in which the future direction of human evolution may be regulated. Three comprehensive review articles by Davis, Lederberg, and English, survey the prospects for, and limitations to, genetic engineering techniques. These review articles are followed by two papers treating amniocentesis, an increasingly used procedure for prenatal detection of many chromosomal and certain gene defects. Muller's classic on "voluntary choice of germ plasm" follows. It illustrates a currently feasible procedure, the application of which is designed to have a beneficial effect on the human gene pool. Muller's paper is followed by two technical papers that deal with procedures that could ultimately permit the correction of certain defects in genetically abnormal humans. Finally this section of the book concludes with articles on "gene surgery" and *in vitro* embryological development. The former procedure, if perfected, would offer hope for correcting specific gene defects in individuals, while both gene surgery and *in vitro* development techniques might be expected to have a potentially beneficial effect on the human gene pool. Before reading these 11 articles, let's briefly survey some of their main themes.

"Prospects for Genetic Intervention in Man" by Bernard D. Davis, Adele Lehman Professor of Bacterial Physiology at Harvard Medical School, is a concise survey of the possibilities, limitations, and dangers associated with various methods of genetic and reproductive engineering. Professor Davis suggests that correcting defects controlled by single genes (e.g., PKU) through replacement of defective genetic material with the "correct" DNA message via transformation or transduction procedures is a much more likely prospect than the genetic manipulation of behavioral traits that are regulated by many genes and strongly influenced by environmental conditions. Davis also briefly discusses cloning, predetermination of sex, and selective breeding as means of genetic engineering. The dangers of various genetic manipulation techniques and their potential misuse for political and military purposes are briefly discussed. Davis concludes by discussing the damage that would be done to society if basic research were curtailed as a consequence of public apprehension over the possible misuse of genetic discoveries.

Nobel laureate in genetics, Joshua Lederberg, of the Stanford University School of Medicine, contrasts the high expectations held for genetic engineering with the reality of our present inability to prevent and treat diseases having a genetic basis. We do not know how to prevent new mutations from occurring; indeed, we usually do not even know that a recessive mutation has arisen until it is segregated in the homozygous condition, perhaps many generations later. In a few cases we have the means to detect the presence of a deleterious recessive gene in heterozygous carriers and, thus, have a basis for counseling the carriers of the probability of their having a homozygous recessive (defective) child. We have amniocentesis, which permits prenatal detection of several chromosome aberrations and a few enzyme deficiencies having a genetic basis. We can treat certain genetic diseases by replacing the missing gene product (e.g., insulin) or by removing from the diet substances (e.g., phenylalanine) that cannot be metabolized. On the whole, however, our arsenal of weapons against genetic disease is quite meager. On the other hand, these inadequacies constitute strong motivation for attempting to devise techniques of genetic manipulation—gene therapy (via transformation or transduction) and cloning being two examples. These inadequacies also increase the risk that genetic manipulation techniques will be used prematurely—before we are fully aware of the technical and ethical problems associated with them. In the case of cloning, Dr. Lederberg specifically notes that "...we simply do not know enough about the question at either a technical or an ethical level....to dogmatize about whether or not it [cloning] should ever be done."

In "Genetic Manipulation and Man," Darrel S. English, associate professor of genetics at Northern Arizona University at Flagstaff, summarizes several techniques that might be used to achieve genetic engineering. He discusses these in the historical perspective of the eugenics movement of the late nineteenth and early twentieth centuries. Dr. English stresses how the contributions made by molecular and microbial geneticists during the last 25 years may play an important role in genetic engineering. He concludes his article with brief comments on the ethical implications of genetic manipulation.

Two articles on amniocentesis summarize the growing importance of this technique in the arsenal of weapons against genetic and chromosomal defects. Paul T. Libassi of the staff of *Laboratory Management* notes that amniocentesis is performed only when the prospective parents agree to an abortion if the fetus is found to be genetically or cytologically defective. At the present time, amniocentesis may be viewed as a technique useful in preventing the birth of a child with a chromosome aberration or one of a few gene-controlled biochemical defects. Because of its relatively limited use (Libassi claims that some 1500 amniocenteses had been performed in the United States as of 1972), the amniocentesis-abortion procedure can be expected to have had virtually no effect to date on the human gene pool. In the second article on amniocentesis, a pioneer in its use, Dr. Henry Nadler, Chairman of the Department of Pediatrics at Northwestern University Medical School, and his colleague, Dr. Albert B. Gerbie, report on how using amniocentesis to monitor 155 "high risk" pregnancies led to the prenatal detection of a total of 13 genetically defective fetuses, including 10 having Down's syndrome (trisomy 21).

Lest the reader conclude that the development of all genetic engineering techniques lies in the future, note Bernard D. Davis' comment that

". . . any society wishing to direct the evolution of its gene pool already has available an alternative approach: selective breeding." The late Nobel laureate, Herman J. Muller, in the sixth article in this section of the book, describes how selective breeding could be accomplished through "voluntary choice of germ plasm." Essentially what Dr. Muller advocated was AID—artificial insemination with the sperm of a donor. Moreover, Muller advocated that the donor be a person of established genetic worth whose contributions to society and whose genetic superiority will have withstood the test of time. Muller further suggested that the semen of outstanding individuals be kept in the deep freeze of a sperm bank until such time that the individual's superiority is clearly recognized. The semen of such individuals could then be used to fertilize the eggs of numerous prospective mothers and thereby increase the frequency of the donor's superior genes in the human gene pool. Although written well over a decade ago, Muller's article continues to have a futuristic ring to it, suggesting that perhaps society is not ready for even this currently feasible method of genetic manipulation.

Two short technical papers—one from *Science* and one from *Nature*—illustrate the continuing search for methods of ameliorating suffering resulting from genetic defects. A. B. Mukherjee and Joseph Krasner of the School of Medicine, State University of New York at Buffalo, describe liver transplantation experiments in genetically deficient rats, which indicate that appropriate transplants can induce the production of a missing liver enzyme. The potential of this procedure for treating a human enzyme deficiency, such as in phenylketonuria, is obvious. Three Oxford University workers (Schwartz, Cook, and Harris) reported in 1971 in *Nature* on the use of somatic cell hybridization to incorporate a gene for a particular enzyme from a chick erythrocyte into a line of mouse fibroblast cells that were genetically deficient for the enzyme. The authors conclude their paper by suggesting how similar techniques might be used to alleviate the phenotypic consequences of a gene-caused enzyme deficiency in humans.

In a short news summary, Daniel Rabovsky, staff writer for *Science,* reports on several recent attempts to use viruses to transport genes from cells of one species to cells of another. The work of Merril, Geier, and Petricciani (1971), for example, demonstrates that a bacterial gene can be inserted into the chromosome complement of a human cell by a phage, and that in its new location the bacterial gene can produce an enzyme previously missing in the human cell. Although done *in vitro,* such experiments suggest that it may not be long before methods will be found for *in vivo* transduction of genes in humans with single gene defects.

Further pursuing the discussion of techniques for genetic manipulation that might be extensively used in the future, Drs. Theodore Friedmann and Richard Roblin (both of whom are affiliated with the Council for Biology in Human Affairs of the Salk Institute for Biological Studies at La Jolla, California) discuss the use of ". . . exogenous 'good' DNA. . . to replace the defective DNA in those [individuals] who suffer from genetic defects." Such gene therapy or surgery might be accomplished by supplying the deficient human cells directly with the desired "good" DNA (transformation) or by having viruses serve as the vector for the introduction of the desired genetic material into the human genome (transduction). The authors propose ethico-scientific criteria that any techniques for gene therapy should satisfy. Note that where Dr. Muller was primarily concerned about using "volutary choice of germ plasm" to modify the human gene pool and thus the evolutionary future of mankind, Drs. Friedman and Roblin direct their concern to alleviating genetic disease in individual patients. There is no doubt, however, that wide scale application of gene therapy techniques could modify the human gene pool. Still, this difference in emphasis could well be the basis for deciding whether or not a genetic engineering procedure is ethical: Is the intervention designed to correct a genetic defect in an individual, or is it designed to modify the evolutionary future of all mankind?

This section of the book is concluded with a discussion by Jean L. Marx, staff writer for *Science,* of recent advances in manipulation of mammalian (including human) reproduction. Marx notes the economic significance of research involving *in vitro* fertilization and the freezing of blastocysts in domesticated animals such as cattle. Finally, a separate section summarizes some of the ethical and legal implications of such research involving human embryos. Although Marx does not specifically mention it, the type of research described is of significance to the cloning of humans, since without the ability to obtain human eggs, to enucleate them, to renucleate them, and to begin embryogeny *in vitro,* cloning human beings would appear to be impossible. In some ways this entire article has been rendered obsolete by the July, 1974 announcement that human eggs fertilized *in vitro* and subsequently implanted in the uteri of receptive females, have been brought to full term and have resulted in what appear to be normal children.

5
Prospects for Genetic Intervention in Man

BERNARD D. DAVIS

Extrapolating from the spectacular successes of molecular genetics, a number of essays and symposia (1) have considered the feasibility of various forms of genetic intervention (2) in man. Some of these statements, and many articles in the popular press, have tended toward exuberant, Promethean predictions of unlimited control and have led the public to expect the blue-printing of human personalities. Most geneticists, however, have had more restrained second thoughts.

Nevertheless, recent alarms about this problem have caused wide public concern, and understandably so. With nuclear energy threatening global catastrophe and with so many other technological advances visibly damaging the quality of life, who would wish to have scientists tampering with man's inner nature? Indeed, fear of such manipulation may arouse even more anxiety than fear of death. The mass media have accordingly welcomed sensational pronouncements about the dangers.

While such dangers clearly exist, it also seems clear that some scientists have dramatized them (3) in order to help persuade the public of the need for radical changes in our form of government (4). But however laudable the desire to improve our social structure, and however urgent the need to improve our protection against harmful uses of science and technology, exaggeration of the dangers from genetics will inevitably contribute to an already distorted public view, which increasingly blames science for our problems and ignores its contributions to our welfare. Indeed, irresponsible hyperbole on the genetic issue has already influenced the funding of

Science, 1970, Vol. 170, Issue 3964, pp. 1279-1283. Copyright © 1970 by the American Association for the Advancement of Science.

research (5). It therefore seems important to try to assess objectively the prospects for modifying the pattern of genes of a human being by various means. But let us first note two genetic principles that must be taken into account.

Relevant Genetic Principles

Polygenic Traits and Behavioral Genetics

The recognition of a gene, in classical genetics, depends on following the distribution of two alternative forms (alleles) from parents to progeny. In the early years of genetics, after the rediscovery of Mendel's laws in 1900, this analysis was possible only for those genes that exerted an all-or-none control over a corresponding monogenic trait—for example, flower color, eye color, or a hereditary disease such as hemophilia. The study of such genes has continued to dominate genetics. However, monogenic traits constitute a small, special class. Most traits are polygenic: that is, they depend on multiple genes, and so they vary continuously rather than in an all-or-none manner. Moreover, each gene itself is polymorphic—that is, it is capable of existing, as a result of mutation, in a variety of different forms (alleles); and though the protein products of these alleles differ only slightly in structure, they often differ markedly in activity.

For our purpose it is especially pertinent that the most interesting human traits—relating to intelligence, temperament, and physical structure—are highly polygenic. Indeed, man undoubtedly has hundreds of thousands of genes for polygenic traits, compared with a few hundred recognizable through their control over monogenic traits. However, the study of polygenic inheritance is still primitive; and the difference from monogenic inheritance has received little public attention. Education on the distinction between monogenic and polygenic inheritance is clearly important if the public is to distinguish between realistic and wild projections for future developments in genetic intervention in man.

Interaction of Heredity and Environment

The study of polygenic inheritance is difficult in part because it requires statistical analysis of the consequences of reassortment, among the progeny, of many interacting genes. In addition, even a full set of relevant genes does not fixedly determine the corresponding trait. Rather, most genes contribute to determining a *range of potential* for a given trait in an individual, while his past and present environments determine his phenotype (that is, his actual state) within that range. At a molecular level the explanation is now clear: the structure of a gene determines the structure of a corresponding protein, while the interaction of the gene with subtle regulatory mechanisms, which respond to stimuli from the environment, determines the amount of the protein made. Hence, the ancient formulation of the question of heredity versus environment (nature versus nurture) in qualitative terms has presented a false dichotomy, which has led only to sterile arguments.

Possibilities in Genetic Manipulation

Somatic Cell Alteration

Bacterial genes can already be isolated (6) and synthesized (7); and while the isolation of human genes still appears to be a formidable task, it may also be accomplished quite soon. We would then be able to synthesize and to modify human genes in the test tube. However, the incorporation of externally supplied genes into human cells is another matter. For while small blocs of genes can be introduced in bacteria, either as naked DNA (transformation) or as part of a nonlethal virus (transduction), we have no basis for estimating how hard it will be to overcome the obstacles to applying these methods to human cells. And if it does become possible to incorporate a desired gene into some cells, in the intact body, incorporation into all the cells that could profit thereby may well remain difficult. It thus seems possible that diseases depending on deficiency of an extracellular product, such as insulin, may be curable long before the bulk of hereditary diseases, where an externally supplied gene can benefit only those defective cells that have incorporated it and can then make the missing cell component.

Such a one-shot cure of a hereditary disease, if possible, would clearly be a major improvement over the current practice of continually supplying a missing gene product, such as insulin. (It could be argued that improving the soma in this way, without altering the germ calls, would help perpetuate hereditary defectives; but so does conventional medical therapy.) The danger of undesired side effects, of course, would have to be evaluated, and the day-to-day medical use of such material would have to be regulated: but these problems do not seem to differ significantly from those encountered with any novel therapeutic agent.

Germ Cell Alteration

Germ cells may prove more amenable than somatic cells to the introduction of DNA, since they could be exposed in the test tube and therefore in a more uniform and controllable manner. Another conceivable approach might be that of *directed mutagenesis:* the use of agents that would bring about a specific desired alteration in the DNA, such as reversal of a mutation that had made a gene defective. So far, however, efforts to find such directive agents have not been successful: all known mutagenic agents cause virtually random mutations, of which the vast majority are harmful rather than helpful. Indeed, before a mutagen could be directed to a particular site it would probably have to be attached first to a molecule that could selectively recognize a particular stretch of DNA (8); hence a highly selective mutagen would have to be at least as complex as the material required for selective genetic recombination.

If predictable genetic alteration of germ cells should become possible it would be even more useful than somatic cure of monogenic diseases, for it could allow an individual with a defective gene to generate his own progeny without condemning them to inherit that gene. Moreover, there would be a longterm evolutionary advantage, since not only the immediate product of the correction but also subsequent generations would be free of the disease.

Genetic Modification of Behavior

In contrast to the cure of specific monogenic diseases, improvement of the highly polygenic behavioral traits would almost certainly require the replacement, in germ cells, of a large but specific complement of DNA. Since I find such replacement, in a controlled manner, very hard to imagine, I suspect that such modifications will remain indefinitely in the realm of science fiction, like the currently popular extrapolation from the transplantation of a kidney or a heart, with a few tubular connections, to that of a brain, with hundreds of thousands of specific neural connections. However, this consideration would not apply to the possibility of impairing cerebral function by genetic transfer, since certain monogenic diseases are known to cause such impairment.

Copying by Asexual Reproduction (cloning)

We now know that all the differentiated somatic cells of an animal (those from muscle, skin, and the like) contain, in their nuclei, the same complete set of genes. Every somatic cell thus contains all the genetic information required for copying the whole organism. In different cells different subsets of genes are active, while the remainder are inactive. Accordingly, if it should become possible to reverse the regulatory mechanism responsible for this differentiation any cell could be used to start an embryo. The individual could then be developed in the uterus of a foster mother, or eventually in a glorified test tube, and would be an exact genetic copy of its single parent. Such asexual reproduction could thus be used to produce individuals of strictly predictable genetic endowment; and there would be no theoretical limit to the size of the resulting clone (that is, the set of identical individuals derivable from a single parent and from successive generations of copies).

Though differentiation is completely reversible in the cells of plants (as in the transfer of cuttings), it is ordinarily quite irreversible in the cells of higher animals. This stability, however, depends on the interaction of the nucleus with the surrounding cytoplasm; and it is now possible to transfer a nucleus, by microsurgery or cell fusion, into the cytoplasm of a different kind of cell. Indeed, in frogs differentiation has been completely reversed in this way: when the nucleus of an egg cell is replaced by a nucleus from an intestinal cell embryonic development of the hybrid cell can produce a genetic replica of the donor of the nucleus (9). This result will probably also be accomplished, and perhaps quite soon, with cells from mammals. Indeed, there is considerable economic incentive to achieve this goal, since the copying of champion livestock could substantially increase food production.

Another type of cloning can already be accomplished in mammals: when the relatively undifferentiated cells of an early mouse embryo are gently separated each can be used to start a new embryo (10). A large set of identical twins can thus be produced. However, they would be copies of an embryo of undetermined genetic structure, rather than of an already known adult. This procedure therefore does not seem tempting in man, unless the production of identical twins (or of greater multiplets) should develop special social values, such as those suggested by Aldous Huxley in *Brave New World.*

Predetermination of Sex

Though no one has yet succeeded in directly controlling sex by separating XX and XY sperm cells, this technical problem should be soluble. Moreover, in principle it is already possible to achieve the same objective indirectly by aborting embryos of the undesired sex: for the sex of the embryo can be diagnosed by tapping the amniotic fluid (amniocentesis) and examining the cells released into that fluid by the embryo.

Wide use of either method might cause a marked imbalance in the sex ratio in the population, which could lead to changes in our present family structure (and might even be welcomed in a world suffering from overpopulation). Alternatively, new social or legal pressures might be developed to avert a threatened imbalance (11). But though there would obviously be novel social problems, I do not think they would strain our powers of social adaptation nearly as much as some urgent present problems.

Selective Reproduction

A discussion of the prospects for molecular and cellular intervention in human heredity would be incomplete without noting that any society wishing to direct the evolution of its gene pool already has available an alternative approach: selective breeding. This application of classical, transmission genetics has been used empirically since Neolithic times, not only in animal husbandry, but also, in various ways (for example, polygamy, *droit de seigneur,* caste system), in certain human cultures. Declaring a moratorium on genetic research, in order to forestall possible future control of our gene pool, would therefore be locking the barn after the horse was stolen.

Having reviewed various technical possibilities, I would now like to comment on the dangers that might be presented by their fulfillment and to compare these with the consequences of efforts to prevent this development.

Evaluation of the Dangers

Gene Transfer

I have presented the view that if we eventually develop the ability to incorporate genes into human germ cells, and thus to repair monogenic defects, we would still be far from specifying highly polygenic behavioral traits. And with somatic cells such an influence seems altogether excluded. For though genes undoubtedly direct in considerable detail the pattern of development of the brain, with its network of connections of 10 billion or more nerve cells, the introduction of new DNA following this development clearly could not redirect the already formed network; neither could we expect it to modify the effect of learning on brain function.

To be sure, since we as yet have little firm knowledge of behavioral genetics we cannot exclude the possibility that a few key genes might play an especially large role in determining various intellectual or artistic potentials or emotional patterns. But even if it should turn out to be technically possible to tailor the psyche significantly by the exchange of a small number of genes in germ cells, it seems extremely improbable that this procedure would be put to practical use. For it will always be much easier, as Lederberg (12) has emphasized, to obtain almost any desired genetic pattern by copying from the enormous store already displayed in nature's catalog.

While the improvement of cerebral function by polygenic transfer thus seems extremely unlikely, one cannot so readily exclude the technical possibility of impairing this function by transfer of a monogenic defect. And having seen genocide in Germany and massive defoliation in Vietnam, we can hardly assume that a high level of civilization provides a guarantee against such an evil use of science. However, several considerations argue against the likelihood that such a future technical possibility would be converted into reality. The most important is that monogenic diseases, involving hormonal imbalance or enzymatic deficiencies, produce gross behavioral defects, whose usefulness to a tyrant is hard to imagine. Moreover, even if gene transfer is achieved in cooperating individuals, an enormous social effort would still be required to extend it, for political or military purposes, to mass populations. Finally, in contrast to the development of nuclear energy, which arose as an extension of already accepted military practices, the potential medical value of gene transfer is much more evident than its military value; hence a "genetic bomb" could hardly be sprung on the public as a secret weapon. Accordingly, we are under no moral obligation to sacrifice genetic advances now in order to forestall such remote dangers: if and when gene transfer in man becomes a reality there would still be time to assert the cultural and medical traditions that would promote its beneficial use and oppose its abuse.

This last obstacle would be eliminated if it should prove possible to develop a virus that could be used to infect a population secretly with specific genes, and it is the prospect of this ultimate horror that seems to cause most concern. However, for reasons that I have presented above the technical possibility of producing useful modifications of personality by infections of germ cells seems extremely remote, and the possibility of doing so by infecting somatic cells in an already developed individual seems altogether excluded. These fears thus do not seem realistic enough to help guide present policy. Nevertheless, the problem cannot be entirely ignored: in a country that has recently been embarrassed by its accumulation of rockets containing nerve gas even the remote possibility of handing viral toys to Dr. Strangelove will require vigilance.

Genetic Copies

If the cloning of mammals becomes technically feasible its extension to man will undoubtedly be very tempting, on the grounds that enrichment for proved talent by this means might enormously enhance our culture, while the risk of harm seemed small. Since society may be faced with the need to make decisions in this area quite soon, I would like to offer a few comments in the hope of encouraging public discussion.

On the one hand, in fields such as mathematics or music, where major achievements are restricted to a few especially gifted people, an increase in their number might be enormously beneficial—either as a continuous supply from one generation to another or as an expanded supply within a generation. On the other hand, a succession of identical geniuses might exert an excessively conservative influence, depriving society of the richness that comes from our inexhaustible supply of new combinations of genes. Or genius might fail to flower, if its drive depended heavily on parental influence or on cultural climate. And in the literary, social, and political areas the cultural climate surely plays so large a role that there may be little basis for expecting outstanding achievement to be continued by a scion. The world might thus be quite disappointed by the contributions of another Tolstoy, Churchill, or Martin Luther King, or even another Newton or Mozart. Moreover, though experience with monozygotic twins is somewhat reassuring, persons produced by copying might suffer from a novel kind of "identify crisis."

Though our system of values clearly places us under moral obligation to do everything possible to cure disease, there is no comparable basis for using cloning to advance culture. The responsibility for initiating such a radical departure in human reproduction would be grave, and surely many will feel that we should not do so. But I suspect that it would be impossible to enforce any such prohibition completely: the potential gain seems too large, and the procedure would require the cooperation of only a very small group of people. Hence whatever the initial social consensus, I suspect that a stable attitude would not emerge until after some early tests, whether legal or illegal, had demonstrated the magnitude of the problems and of the gains.

A much greater threat, I believe, would be the use of cloning for the large-scale amplification of a few selected individuals. Who would wish to send a child to a school with a large set of identical twins as his classmates? Moreover, the success of a species depends not only on its adaptation to its present environment but also on its possession of sufficient genetic variety to include some individuals who could survive in any future environment. Hence if cloning were extended to the point of markedly homogenizing the population, it could create an evolutionary danger. However, we have already lived for a long time with a similar possibility: any male can provide a virtually limitless supply of germ cells, which can be used in artificial insemination; yet genetic homogenization by this means has not become the slightest threat. Since cloning is unlikely to become nearly so easy it is difficult to see a rational basis for the fear that its technical possibility would increase the threat.

Implications for Genetic Research

Though the dangers from genetics seem to me very small compared with the immense potential benefits, they do exist: its applications could conceivable be used unwisely and even malevolently. But such potential abuses cannot be prevented by curtailing genetic research. For one thing, we already have on hand a powerful tool (selective breeding) that could be used to influence the human gene pool, and this technique could be used as wisely or unwisely as any future additional techniques. Moreover, since the greatest fear is that some tyrant might use genetic tools to regulate behavior, and especially to depress human potential, it is important to note that we already have on hand pharmacological, surgical, nutritional, and psychological methods that could generate parallel problems much sooner. Clearly, we shall have to struggle, in a crowded and unsettled world, to prevent such a horrifying misuse of science and to preserve and promote the ideal of universal human dignity. If we succeed in developing suitable controls we can expect to apply them to any later developments in genetics. If we fail—as we may—limitations on the progress of genetics will not help.

If, in panic, our society should curtail fundamental genetic research, we would pay a huge price. We would slow our current progress in recognizing defective genes and preventing their spread; and we would block the possibility of learning to repair genetic defects. The sacrifice would be even greater in the field of cancer: for we are on the threshold of a revoluntionary improvement in the control of these malignant hereditary changes in somatic cells, and this achievement will depend on the same fundamental research that also contributes toward the possibilities of cloning and of gene transfer in man. Finally, it is hardly necessary to note the long and continuing record of nonmedical benefits from genetics, including increased production and improved quality of livestock and crops, steadier production based on resistance to infections, vastly increased yields in antibiotic and other industrial fermentations, and, far from least, the pride that mankind can feel in one of its most imaginative and creative cultural achievements: understanding of some of the most fundamental aspects of our own physical nature and that of the living world around us.

While specific curtailment of genetic research thus seems impossible to justify, we should also consider briefly the broader proposal (see, for example, 8) that we may have to limit the rate of progress of science in general, if we wish to prevent new powers from developing faster than an inadequate institutional framework can be adjusted to handle them. While one can hardly deny that this argument may be valid in the abstract, its application to our present situation seems to me dangerous. No basis is yet in sight for calculating an optimal rate of scientific advance. Moreover, only recently have we become generally aware of the need to assess and control the true social and environmental costs of various uses of technology. Recognition of a problem is the first step toward its solution, and now that we have taken this step it would seem reasonable to assume, until proved otherwise, that further scientific advance can contribute to the solutions faster than it will expand the problems.

Another consideration is that we cannot destroy the knowledge we already have, despite its potential for abuse. Nor can we unlearn the scientific method, which is available for all who wish to wrest secrets from nature. So if we should choose to curtail research in various fundamental areas, out of fear of possible long-range application, we must recognize that other societies may make a different choice. Knowledge is power, and power can be used for good or for evil; and, since the genie that brings new knowledge is already out of the bottle, we must learn to direct the use of the resulting power rather than curse the genie or try to confine him.

We cannot see how far the use of science as a scapegoat for many of our social problems will extend. But the gravity of the threat may be underscored by recalling that another politically based attack on science, Lysenkoism, utterly destroyed genetics in the Soviet Union and seriously crippled agriculture, from 1935 to 1965 (13). [This development illustrates ironically the unstable relation between political and scientific idea: for Karl Marx had unsuccessfully requested permission to dedicate the second volume of *Das Kapital* to Charles Darwin (14)!] Moreover, the current attacks on genetics from the New Left can build on, and have no doubt contributed to, widespread public anxiety concerning gene technology. Thus while a recent report prepared for the American Friends Service Committee (15) presents an open and thoughtful view on such questions as contraception, abortion, and prolongation of the period of dying, it is altogether opposed to any attempted genetic intervention, including the cure of hereditary disease.

Genetics will surely survive the current attacks, just as it survived attacks from the Communist Party in Moscow and from fundamentalists in Tennessee. But meanwhile if we wish to avert the danger of some degree of Lysenkoism in our country we may have to defend vigorously the value of objective and verifiable knowledge, especially when it comes into conflict with political, theological, or sociological dogmas.

References and Notes

1. P. B. Medawar, *The Future of Man* (Basic Books, New York, 1960); Symposium on "Evolution and Man's Progress," *Daedalus* (Summer, 1961); G. Wolstenholme, Ed., *Man and His Future* (Little, Brown, Boston, 1963); J. Lederberg, *Nature* **198**, 428 (1963); J.S. Huxley, *Essays of Humanist* (Harper and Row, New York, 1964); T. M. Sonneborn, Ed., *The Control of Human Heredity and Evolution* (Macmillan, New York, 1965); R. D. Hotchkiss, *J. Hered.* **56**, 197 (1965); J. D. Roslansky, Ed., *Genetics and the Future of Man* (Appleton-Century-Crofts, New York, 1966); N. H. Horowitz, *Perspect. Biol. Med.* **9**, 349, (1966).
2. The term "genetic engineering" seemed at first to be a convenient designation for applied molecular and cellular genetics. However, I agree with J. Lederberg [*The New York Times*, Letters to the editor, 26 September (1970)] that the overtones of this phrase are undesirable.
3. Editorials, *Nature* **224**, 834, 1241 (1969); J. Shapiro, L. Eron, J. Beckwith, *ibid.*, p. 1337.
4. J. Beckwith, *Bacteriol. Rev.* **34**, 222 (1970).
5. P. Handler, *Fed. Proc.* **29**, 1089, (1970).
6. J. Shapiro, L. MacHattie, L. Eron, G. Ihler, K. Ippen, J. Beckwith, *Nature* **224**, 768 (1969).
7. K. L. Agarwal, and 12 others, *Ibid.* **227**, 27, (1970).
8. S. E. Luria, in *The Control of Human Heredity and Evolution*, T. M. Sonneborn, Ed. (Macmillan, New York, 1965), p.1.
9. R. Briggs and T. J. King, in *The Cell*, J. Brachet and A. E. Mirsky, Eds. (Academic Press, New York, 1959), vol. 1; J. B. Gurdon and H. R. Woodward, *Biol. Rev.* **43**, 244 (1968).
10. B. Mintz, *J. Exp. Zool.* **157**, 85, 273 (1964).
11. A. Etzioni, *Science* **161**, 1107 (1968).
12. J. Lederberg, *Amer. Natur.* **100**, 519 (1966).
13. Z. A. Medvedev, *The Rise and Fall of T. D. Lysenko* (Columbia Univ. Press, New York, (1969).

14. T. Dobzhansky, *Mankind Evolving* (Yale Univ. Press, 1962), p. 132.
15. *Who Shall Live* Report prepared for the American Friends Service Committee (Hill and Wang, New York, 1970).

6
Genetic Engineering and the Amelioration of Genetic Defect

JOSHUA LEDERBERG

Few subjects pose as many difficulties for rational discussion as does the bearing of genetic research on human welfare. It is monotonously coupled with such inflammatory themes as racism, the decline of the species, overpopulation, hidden genocide, religious debates on abortion and contraception, the plight of the individual in mass society, and "how many generations of idiots is enough?" Apart from these cliches of social controversy, we also face serious problems in clarifying the technical context in a field where scientific innovations have far outpaced their technical application and applicability to man. Should we spend much time worrying about the ethical implications of the genetic findings of the next century, when we must do this on the basis of a set of assumptions about the human conditions that will surely change dramatically in every other way?

The gap between elementary principle and practical realization poses many dilemmas for a fair-minded evaluation, and a brief commentary cannot do justice to all of the relevant issues. The scientist who concentrates on exposing the technical possibilities will be castigated for ignoring the ethical implications; and if he goes further afield, he will be accused of sermonizing, and indeed may be overreaching his particular area of competence.

According to journalistic accounts, we will shortly be writing prescrip-

Reprinted from *Bio Science* Vol. 20, pp. 1307-1310, December 15, 1970 by permission of the American Institute of Biological Sciences.

tions for human quality to order. "Do you want your baby to be 8 ft tall, or have four hands—just tell the geneticist, and he will arrange it for you," goes this line of advertisement. But the most sophisticated geneticist today is baffled by challenges such as Huntingdon's disease. Will the son of an afflicted father be afflicted later in life? What can he do to assure that his own children will not have it?

Perhaps some year soon we will know enough at least to recognize the genotype before neuronal degeneration has been irreversibly set in motion. But our failure to be able to provide significant help today is a humbling reality next to the effusive but justifiable predictions about future accomplishments.

What then of the bold claims for a brave new world of genetic manipulation? Their substance is grounded on the recent solution of many fundamental mysteries of genetic biochemistry. Many of the obstacles to genetic engineering, apart from the moral and political questions that it may pose, are technological; which is to say that their solution is consistent with our basic scientific knowledge of the gene. But this is as if to say that "merely technical obstacles" prevent building a land bridge from San Francisco to Honolulu. It is safe to predict that this enterprise will never eventuate, not merely because it would be a million times more costly than previous bridges, but rather because other challenges will compete for the energies and resources. The presumed benefits will be achieved by other routes, the image of the trans-Pacific bridge will persist as a metaphor, reminding us of technical achievements in other fields of transportation and communication and of the political prodigy of the evolution of a specific island from dependency to statehood.

Construction works like bridges, are open to evaluation and judgment obvious to all. Biotechnical projects are more likely to be cloaked in an esoteric jargon that depletes common sense. We may then hear the most absurd generalizations, such as "whatever is technically feasible tends to get done."

Anyone who has actually labored to do *anything* knows that the more appropriate slogans are "almost nothing ever gets done, especially if it costs money." Or "when a need is generally perceived, articulately formulated and wisely analyzed, the technical problems will be surmounted. But this will happen much sooner if a mass advertising campaign can be built around it." If transoceanic bridges become fashionable (as might happen incidentally during an arms race), they will be built, and probably the same can be said for genetic engineering. Our foresight about the future will prove to be right or wrong more on the basis of the predictability of fashion than of the scientific bases to technical solutions. I do not venture to foresee the directions that such fashion may take. This is more the realm of the political theorists or of authors like John Brunner (*Stand on Zanzibar*).

Where then does the scientist fit into such a discussion?

He can fairly justify his life and work in terms of fundamental knowledge about nature, Studies on the implantation of nuclei into eggs of different genotypes are a rewarding approach to learning how genes function and how this relates to the development of the egg. Were they done for the purported purpose of learning the technology of cloning in man, we would then be obliged to set a priority (positive or negative) on it from the standpoint of the human values that might justify or repudiate the investment. A small amount of "scientific" effort is, unhappily, biased by the expectation of the publicity that will attach to spectacular demonstrations of "behavior control" or "gene control" (shall we also say "moon-walking?") for its own sake; the scientific community can seek to impose criteria of *scientific* validity for the funding of such projects; or, failing this, to dissociate itself from the responsibility when (as is customary) it does not have the authority to make the critical decisions.

Alternatively, the scientist can function as the actual or effective member of a technological team that will address itself to the solution of grave problems that encumber human welfare. Then we must (and usually do) insist that the problems are real ones, and that technical solutions are credible. What is more often obscured is the need to examine all the side effects, to inhibit the premature exploitation of new cures that may be far worse than the disease, to assure that as much sophistication goes into looking for the side effects as was eagerly purchased for the primary solution. The hazards of suboptimal solutions are well appreciated for drugs, but we are just now feeling their full force in such disparate fields as pesticides and auto transport. Pesticide poisoning and air pollution have been figured as technological jinn. It would be more fair to lay the blame on technological idiocy—the refusal to make the economic investments needed to develop all the science required for the safe and healthful utilization of the new tricks.

What then are the *problems* to which genetic science can be applied? Some may think of rescuing man from the prospect of nuclear annihilation by recasting the genes for aggression, or acquiescence, that are supposed to predestine a future of territorial conflict. Even if we postulate for the sake of agreement that we knew the genetics of militarism, we have no way to apply it without solving the political problem that is the primary difficulty. If we could agree upon applying genetic (or any other effective) remedies to global problems in the first place, we probably would need no recourse to them in the actual event.

The converse argument applies to the gloomier predictions of totalitarian abuse of genetic technology. The scenario of *Brave New World* is well adertised by now, and no one doubts that a modern slave state would reinforce its class stratification by genetic controls. But it could not do so without having instituted slavery in the first place, for which the control of the mass media presents much more immediate dangers than knowledge of DNA. It is indeed true that I might fear the control of my behavior through electrical impulses directed into my brain, but (possibly excepting television) I do not accept the implantation of the electrodes except at the point of a gun, and this is the problem.

So much for the grand designs of genetic engineering. There remain the very real tragedies of genetic disease. The societal interest in preventing or ameliorating mental retardation and other forms of congenital malformation is obvious. (The true cost of lifetime maintenance of a 21-trisomy approaches a megadollar.) It is also entirely congruent with the needs of the family, and, if we believe in the nobility of man and the worth of *human* life, also of the afflicted child as well.

The most effective avenues of preventing genetic disease include (1) the primary prevention of gene mutations, and (2) the detection and humane containment of the DNA lesions once introduced into the gene pool. The "natural" mutation process in man results in the introduction of a new bit of genetic misinformation once in every ten gametes. Most of the human cost of this "mutational load" is paid during early stages of fertilization and pregnancy, where it makes up a fair part of the total fetal wastage. But about 2% of newborns suffer from a recognizable discrete genetic defect. This is just the tip of the iceberg; the heritability of many common diseases suggests that from one-quarter to one-half of all disease is of genetic origin, for there are important variations in susceptibility to the frankest of environmental insults.

Not all of this health deficit can be attributed to recurrent mutations. An unknown proportion results from the selective advantage that is paradoxically associated with the heterozygous state of many genes, even some with lethal effect in the homozygote (like sickle-cell hemoglobin). Nevertheless, a significant part of medicine—much more than most practitioners overtly recognize—is in fact directed to lesions that are inherently preventable if we could control the mutation process in the background.

About a tenth of the "natural mutation rate" can be attributed to background radiation from cosmic rays and from radioactive potassium and other isotopes in our natural environment. Therefore, doubling the background, which would correspond to the maximum permissible standards now advocated by federal agencies, would add another 10% to the existing mutation rate: 1/9 rather than·1/10 of our gametes would carry deleterious mutations. This is an enormous impact in absolute terms, a modest increase in relative terms. We must nevertheless pay careful attention to the benefits that would be connected with this level of radiation exposure to be sure we are getting a fair bargain.

It must be pointed out that industrial nuclear energy activities today add less than 10% to the average radiation background (hence less than 1% to the mutation rate); medical X-rays add 50% and 5% respectively. The more prevalent standards for the judicious and cost-effective use of diagnostic X-rays do not necessarily or automatically excuse the dispensable residue.

A signficant portion of "spontaneous" mutations must be attributed to environmental chemicals, many of which are clearly established as mutagens in laboratory experiments (for example, the peroxy-compounds that characterize smog). The extent to which such materials reach the germ cells is absolutely unknown at present. However, there are good reasons to believe that (1) the induction of mutations, in germ cells, and of cancer, in somatic cells, are fundamentally similar processes—most chemical carcinogens being also mutagenic when properly tested; and (2) a large part of the incidence of cancer is of chemical-environmental origin, cigarette-smoking being only the best known and best-advertised example. It therefore follows that environmental mutagenesis is equally prevalent, if the relative effects of radiation in the two systems are any hint, the cryptic penalties of the mutations are likely, in the long run, to exact the larger price in human misery.

The direct observation of human populations for evidence of changes in mutation rates is an almost hopeless task. We have no way of managing the tangle of known and unknown environmental influences that bear on different individuals. Nor do we have tractable assays for the occurence of new mutations, whose manifestation may be delayed (by transmission through heterozygotes) for many generations, or confused with malformations due to pre-existing mutant genes, or to non-genetic causes. If we had to rely upon epidemiological evidence, we would still lack persuasive evidence that radiation, even in barely sublethal doses, was mutagenic in man.

Our only recourse is the laboratory experiment with convenient mammals such as the mice, and sometimes even more efficiently with viruses and microorganisms. Even so, only the most potent mutagens can be identified with mice, and many uncertainties will remain that cannot be resolved given possible differences in metabolism and transport, cell selection, intrinsic sensitivity, and the duration and style of life of the human versus the experimental species. Very recently, we have been able to look deeper into the mechanism of chemical mutation, at the level of the structure and repair of the DNA molecule, and new procedures may be developed that can give us more reliable information on the susceptibility of the DNA of human cells to environmental insults. They may also give us clues to ways of neutralizing mutational lesions, either by blocking the primary effects of mutagens on DNA, or by bolstering the natural mechanisms of "editing" and repair of DNA information.

Once a mutation has been allowed to occur in a gamete, and this then participates in fertilization and the production of a new individual, we face a much more difficult problem in any effort at genetic hygiene. For now we must deal with the destinies of human individuals, not merely the chemistry of an isolated segment of DNA. Our problem, seen in the whole, is compounded by every humanitarian effort to compensate for a genetic defect, insofar as this shelters the carrier from natural selection. So it must be accepted that medicine, even prenatal care (which may permit the fragile fetus to survive), already intrudes on the question "Who shall live?", the challenge so often thrust at rational discussions of policies that might influence the frequencies of deleterious genes. It is so difficult to do only good in such matters that we are better off putting our strongest efforts into the prevention of mutations, so as to minimize the heavy moral and other burdens of decision once the gene pool has been seeded with them.

We still cannot evade an evoluntionary legacy of genetic damage that would remain with us for generations, even if all new mutation could be stopped by fiat. Our fundamental resources remain very feeble. In a few cases, we can diagnose the heterozygous carriers of recessive mutations, and the genetic counselor can then advise the prospective parents of the odds that they will have affected children. Where voluntary childlessness is unacceptable, it is also sometimes possible to monitor a pregnancy by sampling cells from the amniotic fluid. This can enable the mother to proceed with confidence or to request an elective abortion on the basis of firm knowledge of the genotype of the fetus. We can expect a rapid extension of technical facilities for such diagnoses. At present, they are limited to examination of the chromosomes (for gross chromosomal abnormalities, like Downs' syndrome), and to enzyme assays on cultured cells, which can diagnose a few dozen rare diseases with varying degrees of reliability. We will surely be learning during the next decade how to use much more sophisticated approaches to the structure of the DNA and RNA of such cells for more basis diagnostic methods.

In many cases, a deeper understanding of the casual chain by which a DNA alteration leads to pathology may help us devise new forms of therapy to compensate for the genetic defect. This may be as crude as the use of insulin in diabetes or as subtle as the use of controlled diets in phenylketonuria. (Both approaches are valuable; neither entirely satisfactory.)

Another approach to constructive therapy, which may mitigate a variety of diseases, is an extension of the existing uses of specific virus strains. At present, their role in medicine is confined to their use as vaccines, for the provocation of immunity against related, wild viruses. This a a specialized example of the modification of cell metabolism by inoculated DNA, discovered empirically by Jenner, and still quite imperfectly understood (our ignorance being concealed by the conceptualizations of clinical virology, which still fail to explain just how a vaccine works—e.g., to state just which cells of the vaccinated individual are carrying the viral genetic information, and in what form. We can visualize the engineering of other viruses so that they will introduce compensatory genetic information into the appropriate somatic cells to restore functions that are blanked out in a given genetic defect. As with vaccine viruses, this presumably will leave the germ-cell DNA unaltered, and therefore does not attack the defective gene as such. If we can cope with the disease, should we bother about the gene? Or may we not leave that problem to another generation.

There has been much to-do about another theoretical possibility, "cloning" a man, as might be done by the renucleation of a fertilized egg with a somatic cell nucleus from an existing individual. Similar experiments have been successfully completed with frogs, and are being attempted with mice. Such experiments with laboratory animals will surely be very fruitful if the technique can be developed. It would also have enormous value in livestock breeding, just as cloning (propagation by cuttings) is a mainstay of horticulture. Until such experiments have been pursued in some depth with other animals, it is merely a speculative game to discuss applying such reproductive novelties to man. There is no urgent social problem to be addressed by such a technique. It does serve as a metaphor to indicate that future generations will have infinitely more powerful ways than we do to deal with whatever they may regard as socially urgent issues of human nature. We can therefore focus more confidently on dealing with the distress of individual human beings in the immediate generation. The metaphor also suggests that intrusive genetic engineering, if it is pursued for any other reason, will have plenty of policy problems to digest even before the "technology" has reached the point of detailed synthesis of genotypes by design.

Finally, medical scientists in general are fully aware and have fully assimilated ethical concerns about the application of new techniques in man, by contrast to experimental animals. For a long time, it has been known that one could operate on the brain in such "interesting" ways as dividing the corpus callosum with the possibility of the development of autonomous "intellects" in the two hemispheres. It would be unthinkable to apply such surgical technology to man without the persuasion and conviction that it would be for the benefit of the patient-subject. We will not be given the benefit of the doubt in public discussions of such questions; there are many influential people who really believe that "anything feasible will be done," and we may have to restate the obvious many times in reviewing the ethical constraints on possible experimentation.

To return to the "clone-a-man" metaphor: in my view, *we simply do not know enough about the question at either a technical or ethical level (and these are intertwined) to dogmatize about whether or not it should ever be done.* Certainly it cannot be thought of within the framework of our generally accepted standards of medical ethics, unless (1) we can make and communicate a reasonably confident prediction of the outcome, and, more important (2) it has the informed consent and serves a reasonable humanitarian purpose for the individuals who are involved. In genetic matters, this must include the interest of the prospective newborn as well as of his parents and of the community. If we demand that he be represented in person, then no one could reasonably be allowed to be born whether by "natural" sexual fertilization by the design of his parents or otherwise. Cloning-a-man is one of the least important questions I can think of; "who must be held accountable for the next generation and how" may be the most important question.

7
Genetic Manipulation and Man

DARREL S. ENGLISH

The science of improving human beings by applying the principles of inheritance to obtain a desirable combination of physical characteristics and mental traits is called eugenics. The term was coined by Francis Galton in 1883; literally translated, it means to be "true born" or "well born."

Although most writers on this subject begin by citing the Greeks, the idea of improving the human stock probably goes back even farther. Even though he lacked any knowledge of the laws of heredity, primitive man could see that parents with imperfections often bore children with the same deficiencies. Perhaps the earliest aim was to produce a race of physically perfect men, capable of coping most efficiently with the tremendous hardships they had to endure, including contests with enemies and wild beasts. Paralleling the desire to develop physically was the need to develop intellectually; and cultures may have tended to develop their mental capacities more than their physiques.

Plato advanced the idea of race improvement by methods similar to those of present-day stock breeders. In *The Republic* he proposed that matings between the most nearly perfect men and women be encouraged and that their offspring be raised in a state nursery. Inferior persons should be prevented from reproducing; and if by chance they should have children these should be destroyed. To some extent Plato's eugenic

Reprinted, with permission, from *American Biology Teacher*, Vol. 34, No. 9.

methods were practiced in Sparta; the result was a population of people with fine physiques (Castle, 1925; Fasten, 1935).

In Athens the emphasis was on art, politics, and science. Here, too, people of good background were encouraged to marry among their kind. Did this pay off? Galton (1909) noted that during the 6th to 4th centuries B.C. Athens produced some of the most illustrious men the world has ever known. Whether the decline of Greek society was due to master—slave intermarriage, as some have suggested, or whether the "inferior" classes reproduced more rapidly than the "superior" classes, one can only guess. Possibly the downfall resulted from economic rather than genetic changes.

The eugenic movement appears to have made little headway until the 19th century. Charles Darwin, Francis Galton, and Gregor Mendel were among the scientists who kindled the spark of modern eugenics; directly or indirectly, their work stimulated interest in this field. The Darwinian concept of the "survival of the fittest" brought to the fore many inescapable implications. In *The Descent of Man* (1874) Darwin wrote: "It is our natural prejudice and arrogance which made our forefathers declare that they were descended from demi-gods and which leads us to demure to this conclusion." The growing evidence supporting the theory of evolution, together with the refinement of man's ability to influence the evolution of domesticated plants and animals, stimulated work along eugenic lines.

Darwin's cousin Francis Galton, the English anthropometrist and examiner of family records, led the first big surge toward eugenic studies. In his book *Enquiries Into Human Faculty and Its Development* (1883) he coined the term eugenics, defining it as the study of agencies, under social control, that could improve or impair the hereditary qualities of future generations, either physically or mentally (to paraphrase the 2nd ed., 1908). He proposed improvement in human breeding by decreasing the birthrate of unfit persons and increasing the birthrate of fit persons. He made extensive studies on criminality, insanity, blindness, and other human defects. Galton was able to understand the inheritance pattern of some human traits. He recognized the importance of twin studies for human genetics and was aware of the social implications of genetic change in man. He was instrumental in applying more sophisticated statistical methods of solving problems of genetics. (It is interesting to note that Galton, who was unusually gifted and was devoted to the principle that better-qualified people should produce at least their share of children, himself died childless.)

Mendel shed new light on the genetics of man with the discovery of fundamental principles of inheritance. Thanks in part to his work, scientists learned how to deal with questions of human heredity in a methodical manner.

Eugenic Implementation

Eugenics is often divided into "negative," or preventive, and "positive," or progressive, eugenics. The first is concerned with the elimination of alleles that produce undesirable phenotypes; the second is concerned with furthering the increase of alleles that produce desirable phenotypes or at least with guarding against the decrease of these alleles. In a sense, the two branches of eugenics are identical: to discourage reproduction among people having undesirable traits is, ipso facto, to encourage a comparatively higher rate of reproduction among people who lack these undesirable traits.

In the past, negative eugenics meant sterilization and institutionalization. The results of these measures were insignificant. In many cases, determining what should be considered undesirable was a major problem; furthermore, procedures were not always ethically acceptable to the majority of the people.

More recent methods of negative eugenics include dissuasion from procreation, voluntary sterilization, medically induced abortions, education as to the genetic basis of human traits, and the encouragement of birth-control practices by persons possessing undesirable traits. These efforts are intended to reduce the dysgenic, or deteriorating, effect on society caused by the perpetuation of certain undesirable traits.

Some biochemical defects, which are known to be propagated by a single defective gene, tend to eliminate themselves naturally; many, however, do not. Hemophilia, for example, has severe and often lethal effects generation after generation. In agammaglobulinemia, children are born without the ability to manufacture antibodies. In phenylketonuria (PKU), children are unable to metabolize phenylalanine; they become mentally incompetent if not treated soon after birth. In certain conditions the genetic defect may not be discovered until after reproductive age has been reached and children have already been introduced into the population. Huntington's chorea, with its progressive deterioration of the muscular and nervous systems, does not make itself known until the victim is in his forties. As a result of this dominant lethal gene, approximately 50% of the offspring can expect to succumb to the same fate.

Positive eugenics, aiming at the reproduction of persons of presumedly superior genotypes, has as many problems associated with it as negative eugenics. This concept came into disrepute because of early notions of who was "desirable" and the classification of certain kinds of people as "degenerates." A major setback for eugenicists came during World War II, when Hilter's eugenic movement went to the extreme of trying to achieve a "master race" of "Aryans" at the expense of "non-Aryans." It is understandable why the term eugenics has some very bad connotations in the minds of many people.

Unfortunately, many characteristics that we might consider desirable—high intelligence, good physical health, longevity—are not under the control of a single genetic factor; instead, they arise from a complex of genes that interact in an appropriate environment. H. J. Muller, J. F. Crow, and others have pointed out that such traits as high intelligence and esthetic sensibility have not been selected with any effort in the past. They postulate, however, that if selection for such traits were to be instituted, the general population might respond very rapidly. The means of selection remains the paramount problem. Selection schemes, even though they might be extremely successful, may prove to be completely intolerable.

During the early 1960s, Muller hotly advocated the use of sperm banks in preference to selective-mating programs. These banks would preserve, frozen, the sperm of men of outstanding qualities. This method, called germinal choice, or eutelegenesis, assumes that married women, otherwise barren, might choose to be artificially impregnated with semen from men who had highly desirable traits. This would be possible even though the donor had died several years before conception took place. Full information about the donor would be provided to ensure the best possible combination of genes (Carlson, 1972).

The possibility of such a "preadoption" method has had some acceptance in the United States. It is estimated that 10,000 artificially inseminated conceptions occur in the United States every year (Taylor, 1968). The reason for most births of this kind is that the husband is impotent or possesses some genetic incompatibility, such as the Rh factor, or harbors a known genetic defect, such as hemophilia.

Euphenics

Molecular biologists and medical researchers are developing a series of procedures for the relief of genetic disorders. The field of euphenics is concerned with the improvement of genotypic maladjustments by treatment of genetically defective persons at some time in their life cycle. Today there are many sensitive tests for genetic defects. These enhance the reliability of counseling and decision-making before or during childbearing. As for treatment: in phenylketoneuria, for example, the child is given the "diaper test," which depends on a color change of the urine when ferric chloride is added; or the Guthrie test, which is based on the ability of certain strains of bacteria to grow on substrates containing high levels of phenylalanine. If a problem exists in the metabolism of phenylalanine and the condition is diagnosed early, the infant is put on a diet low in phenylalanine for the first five years of life, and the brain develops normally.

Another example of technologic success in the detection of genetic abnormality is that of a woman who possessed the potential to produce a mongoloid child. The woman had three sisters who suffered from Down's syndrome, or mongolism, and she feared that her own children might have this condition. In 1959 it was discovered that the syndrome is due to an additional (21st) chromosome, which originates by a mistake in cell division just before conception. The woman requested a study of her cells, and her fears were confirmed: she had the extra chromosome. Genetic counselors advised her that she had one chance in three of producing an abnormal child. Several years ago the woman became pregnant. Doctors informed her that a new technique, called amniocentesis, would enable them to determine whether the fetus was aberrant. The method consists in tapping the fluid of the amnionic sac and making chromosome studies of the cells from the fluid. During the 14th week of pregnancy the woman's fears were borne out: she was told her baby was mongoloid. The pregnancy was terminated by therapeutic abortion. Several months later she became pregnant a second time; once again the tests showed Down's syndrome, and the pregnancy was terminated. There was a third pregnancy; and this time the chromosome studies indicated the baby would be normal—and a boy. The woman at last gave birth to what she had wanted for so long: a normal son. A year and a half later, following carful testing, she gave birth to a normal daughter.

Henry Nadler has used amniocentesis in the diagnosis of high-risk mongolism cases with a high degree of accuracy. This test can be performed between the 12th and 18th weeks of pregnancy. Although there is some danger, the benefits are said to more than justify the risks. Over 35 human enzymatic diseases have been identified by this technique; they include cystic fibrosis, cystinosis, amaurotic idiocy, gout, Gaucher's disease, galactosemia, xeroderma pigmentosum, and diabetes mellitus.

Recent advances in the detection of carriers of recessive diseases, such as hemophilia and some forms of muscular dystrophy, are helping to make the job of genetic counseling an easier one (Friedmann, 1971; Nadler and Gerbie, 1971; *Time,* 1971).

Human genetic analysis of single-gene effects by pedigree analysis is still valid and useful. McKusick (1970) listed 1,487 human traits known to be controlled by a single dominant gene, 531 by a recessive gene, and 119 by X-linked genes. With the advent of computerized technology, experiments may now be designed for determining the genetic mechanisms from family-history data, and additional information will be rapidly added to the catalogue of human genetic defects.

From the viewpoint of the population geneticist, the symptoms of the hereditary diseases may have been treated, but the genes remain unchanged and can be passed on to subsequent generations. Therefore the real genetic problem is not solved and, in fact, such medical practices only compound future problems: by preserving defective genotypes and allowing them to reproduce and transmit these genes, we create a population that is more dependent upon surgery, drugs, and similar treatments.

Furthermore, to simply prevent people who possess defective genes from reproducing will not solve the problem. It is estimated that, on average, each person carries four to eight defective genes that in combination with other defective alleles could bereft a child of his normal faculties. In other words, each conception involves some risk of producing a child with a serious abnormality.

Algeny

The eugenics procedures of the future may be quite different from those of the past. It now appears that the techniques may be at hand to not only "improve" the genetic material by selection but also to correct misinformation within the DNA molecule and thus eliminate the problem.

Genetic engineering, or algeny, is becoming a household word among molecular biologists. Terms such as gene surgery, gene insertion, and gene deletion are beginning to have real meaning for the future of man. The ability to manipulate, in a purposeful manner, the genetic constitution of human beings may come sooner than we anticipate. Taylor (1968) asserted that biologists have reached the critical point of sudden acceleration—the point that physicists reached only a generation ago. It is hoped that biologists and laymen alike will face the problems and potentials of human reproductive engineering in a more realistic and relevant manner than was taken by the atomic physicist and the politicians at the creation of the world's most destructive force (Heim, 1972).

Algeny is a more sophisticated and direct method of curing genetic ills. It is also a permanent cure: in certain instances it might alleviate suffering in future generations. Two approaches seem possible:

1. One might consider the incorporation of a normal gene within the protein coat of some human virus. The cells, after infection with the virus, would supply the defective cell with a corrected copy of the needed information. Although this would only cure the individual during his lifetime and not have any effect on his progeny (unless there might happen to be accidental gene-incorporation), it would be a more direct attack on the genetic problem.

This approach does not appear to be out of the question—as was accidentally demonstrated in a group of laboratory technicians and doctors working with Shope papilloma virus. This virus is capable of inducing tumors in rabbits. Although nonpathogenic in humans, it does infect those who handle it. It appears that one of the viral genes, which is responsible for the synthesis of arginase, is active in human cells. Years after any contact with the laboratory, infected workers were shown to have especially low levels of arginine in their blood as a result of the activity of the viral enzyme (Taylor, 1968). And now genetic engineers are making their first attempts at using a related means of treating this metabolic disease. Investigators at the Oak Ridge National Laboratories and in Cologne are cooperating in the treatment of two German girls who are suffering from low levels of arginase in their blood. It is hoped that injections of live Shope papilloma virus carrying the enxyme will supplement their low levels of arginase. If this is successful, the girls will be able to produce the needed amino acid, arginine, and so achieve a more nearly normal metabolic balance. Someday it may even be possible to produce viruses artificially to correct specific metabolic deficiencies (Gardner, 1972).

2. A second approach to genetic engineering involves replacement of a mutant gene with a normal one by treating germ cells before fertilization. This curative procedure would be even more direct and would have the advantage of being permanent, because descendants would be normal with respect to the defect in question.

In the microbiologic world two methods of incorporating normal genetic material into a mutant cell have been perfected. Avery, MacCleod, and McCarty (1944) exposed nonvirulent bacteria to purified DNA of a virulent strain of bacteria. To their amazement the nonvirulent bacteria were transformed into virulent forms and were able to transmit the trait to future generations. This process is called transformation.

In 1959 three French workers reported that they had extracted DNA from one strain of ducks, called Campbells, and injected the material into a second strain, Pekins. They had hoped to change the offspring somehow, but to their astonishment the injected Pekin ducks began to take on some of the characteristics of the Campbells. This stimulated a flurry of experiments with these strains of ducks, as well as other animals. But, to the dismay of the researchers, subsequent attempts were a total failure, even with the ducks. Not until 1966 was an actual case of transformation verified in organisms other than bacteria. A. S. Fox and S. B. Yoons, of the University of Wisconsin, treated one strain of fruit flies with DNA extracts of another strain; some of the resulting offspring developed genetic anomalies that persisted for several generations (Taylor, 1968).

In contrast with transformation is another mode of permanent genetic change effected by using viruses. This process, called transduction, involves a viral particle that picks up a host gene and, on reinfection in a second host, gives up the genetic material to its new host genome. After incorporation of the newly introduced genome the cell is permanently altered and transmits the acquired trait in the typical genetic manner. The Ukrainian scientist Serge Gerhenson claimed to have transduced a silkworm, using a virus to introduce the foreign DNA. Likewise, there are lines of evidence suggesting that other investigators have been able to transfer drug resistance from one line of mouse-cell cultures to another in this manner (Taylor, 1968).

More recently an exciting experiment was carried out by the molecular biologist Carl Merril and his colleagues: the first successful transplant of bacterial genes into living human cells. Cells from a victim of the genetic disease galactosemia were cultured in vitro. These cells are unable to produce an essential enzyme for the breakdown of the simple sugar galactose. Newborn infants with this defect face malnutrition, mental retardation, and death unless they quickly receive a milk-free diet. Using viruses that had picked up the genes for galactose metabolism from the common intestinal bacterium *Escherichia coli*, the researchers hoped to transmit the gene to human cells in tissue culture. Further investigations showed that they were producing messenger RNA for the missing enzyme and the enzymes themselves. This clearly implied that the cells were being directed to produce the essential enzyme; and equally exciting was the fact that the enzyme-making capabilities were being transmitted to future generations of cells (Merril, Geier, and Petricciani, 1971). These investigators are now striving to make the same kind of genetic transplants with laboratory animals.

These results are particularly significant because they show that bacterial genes can become biologically active in mammalian cells. Furthermore, they clearly establish the universality of the genetic code. And they could have some important implications for the cure of cancerous conditions produced by "runaway" genes. The field of genetic engineering is indeed wide open. It is enough to make one wonder, though, what new genes a person may pick up from the sneeze of the person sitting next to him!

Clonal Reproduction

Once a highly desirable genotype has been produced, it would be beneficial if more of the same organism could be produced. Asexual reproduction of organisms, including man, is another method that may be used someday to produce desirable genotypes. Cloning is the process of inducing normal somatic cells to repeat the complex step in embryo-genesis and eventually produce carbon copies of the original donor organism.

F. C. Steward first showed the feasibility of cloning by taking certain cells from a carrot root and culturing them in coconut milk. Some of these cells formed clumps, which began to differentiate. Transferred to soil, they matured into normal carrot plants. Later experiments have shown that almost any early embryonic cell of the carrot can grow vegetatively (Steward, Mapes, and Smith, 1958).

The possibilities of animal cloning took on reality when J. B. Gurdon, of Oxford University, managed to get the nucleus of an intestinal cell of the South African clawed toad, *Xenopus laevis*, to direct embryogenesis in the enucleated cytoplasm of an unfertilized egg cell. The egg, thus, contained the diploid set of chromosomes and responded by dividing repeatedly. The resulting tadpole was a genetic twin of the toad that had provided the nucleus. By making numerous subclones Gurdon was able to produce many identical copies of the parent toad (Gurdon, 1968). More recently he was able to prepare frogs from cultured cell nuclei (Gurdon, 1970). It would seem there might be no end to the number of copies possible from a single individual. These experiments prove that all the genetic information necessary to produce an organism is encoded in the nucleus of every cell of the organism.

The implications of this technique are numerous. It might be possible to clone a group of Einsteins or Beethovens. Or one might desire a team of astronauts with particular talents and temperaments for a long space voyage. One might be able to clone people with acute psychic awareness, so that extrasensory perception would become commonplace.

A further advantage of this technique has to do with immunologic properties: the members of a clone would be able to accept grafts and tissue or organ transplants without any of the usual repercussions. (The recipient may, however, have some difficulty in convincing one of his clonal twins to give up his heart, lungs, kidneys, or limbs!)

A more practical use of cloning would be its use as a means of tracking extremely deleterious genes during early embryologic development. Someday it may be possible to take the fertilized egg or embryo and culture it in a test-tube. A few cells could be removed and cultured in sufficient numbers for biochemical analysis. If the embryo proved to possess the deleterious genotype, the culture could be terminated; if not, the egg or embryo could be reimplanted in the womb, where development would proceed normally and without further interruption.

Taylor (1968) noted that if vegetative reproduction of human beings is ever achieved, it is most likely to be done by growing a few cells taken from the embryo. The more specialized a cell becomes, the greater the loss of its totipotency. To induce specialized cells, such as nerve, muscle, or brain cells, to become unspecialized once again, appears to be an extremely difficult task.

Of more immediate use to the economy of the world is what one might call a variation on the theme of cloning: the phenomenon of artificial inovulation. All animals seem to be capable of producing many more eggs than they will ever release normally. Injections of the follicle-stimulating hormone (FSH) can induce as many as 40 or more eggs at one time in a cow. This is called superovulation. If the eggs are then artificially fertilized and implanted in a number of competent females, the number of offspring can be multiplied manyfold annually. This process is the converse of artificial insemination, which enables a prize bull to sire more than 50,000 calves a year.

In 1962 two South African ewes gave birth to two lambs whose real parents lived in England. The fertilized eggs had been implanted in the oviduct of a live rabbit, which was flown to South Africa; there the eggs were implanted in the foster ewes. Since then, eggs have been flown to the United States successfully, and transfers between different strains of animals have been accomplished (Taylor, 1968).

Experiments of this kind promise valuable improvements of livestock worldwide. Artificial inovulation also offers hope to those humans afflicted with certain types of sterility. Perhaps the day will come when a woman who would normally be childless will have a prenatally adopted child implanted in her womb, and she will be able to experience all of the emotions and physiologic changes associated with motherhood.

Also related to the ability of man to manipulate the activity of a cell is the phenomenon of regeneration. Mammals have generally lost their totipotency except in specific parts, such as the liver, lymphoid tissue, skin, and bones. The initiatory and regulative factors of regenerative growth are generally unknown. If, however, one could reactivate the genetic events involved in the embryonic organization of cells at the stump of a lost limb, might it not be possible to regenerate the entire structure?

Robert O. Becker, of the Veterans Administration Hospital in Syracuse, N.Y., has said that the possibility of regenerating limbs of mammals, including man, seems nearer. He has successfully stimulated partial regeneration of limbs in rats by the induction of a blastema in response to minute electrical currents applied to the severed area (Becker, 1972). The production of several centimeters of regenerative growth in experimental mammals suggests that higher animals do have regenerative potential if the cells involved can be induced to take on the more primitive state of development. Eventual total regeneration of organs and limbs would make the immunologic complications in transplants a thing of the past.

Conclusion

At one time artificial insemination in humans, sex determination before birth, sex reversal, embryonic determination of genetic aberrations, and organ transplants were looked upon as remote possibilities. Today they are facts; and society has, to some degree, accepted them.

Will man be able to alter his own makeup so profoundly that he will be essentially a new species? Undoubtedly he will strive, with even more enthusiasm, to unlock more of nature's secrets. But some scientists see far greater dangers in man's new knowledge than may be envisioned at first. Seymour Kessler, of Stanford University, has said he "would hate to see manipualtion of genes for behavioral ends because as man's environment changes and as man changes his environment, it is important to maintain flexibility" (quoted in Taylor, 1968). One must be cautious about following a path that eliminates variability, because without it we could go the way of the dinosaurs. This is a particular danger if cloning should become popular. Thus, with increased knowledge comes the responsibility to use this new information wisely (Heim, 1972).

Many biologists are concerned over the moral, ethical, and social implications of the new biology. Although predictions of human genetic control are probably premature—many of the techniques are far from perfected—it is only a matter of time before test-tube babies, cloning of humans, and corrective gene-surgery will be possible.

Numerous national and regional commissions have been set up to study some of the problems. Joshua Lederberg, of Stanford University, does not believe that the perfect human being is a proper goal for the molecular geneticist, even if the techniques could be perfected. Nevertheless, there are those who disagree. Who, then, decides what qualities are to be preserved and by whose standards are they to be directed? J. D. Watson, in an article entitled, "Moving toward the Clonal Man: Is This What You Want?" said that he hopes the mankind will throughly discuss these issues in the next decade. On the other hand, Glenn T. Seaborg, former director of the Atomic Energy Commission, has stated that he believes decisions should be made by experts without the benefit of public debate. The latter comment reminds us of the warning of the late C. S. Lewis, over a quarter of a century ago: "Man's power over Nature is really the power of some men over other men, with Nature as their instrument." One may agree with Senator Walter F. Mondale, who said: "There may still be time to establish some ground rules." And one may doubt whether we really need a clone of people or the creation of a laboratory full of orphaned test-tube babies at a time when the population explosion is one of man's greatest biologic problems. (For citations of views in this paragraph see Taylor, 1968, and Wallace, 1972.)

And yet . . . genetic engineering does promise to alleviate many of man's ills and sufferings. What a relief from anxiety it would be to know that a genetic disease you might harbor would not be passed on to your child! Can one begin to express the gratitude of a mother as she watches her child, who might have been mentally retarded a few years ago, present the valedictory speech at her high school graduation? Amniocentesis is likely to show enormous cost benefits in the treatment of Down's syndrome in the near future: 4,000 mongoloid infants are born each year in the U.S.; the lifetime institutional care for each one is approximately $250,000; and therefore, unless women carrying such abnormal fetuses are encouraged to have therepeutic abortions, their care will cost society some $1.75 billion nationally by 1975 (Friedmann, 1971; *Time,* 1971).

One can see that man faces many new legal, moral, and ethical questions. Will he gave to consider assigning certain rights to the fetus? At what point does a fetus have legal rights, and to whom does the test-tube baby turn for support and legal inheritance? What will be the nature of the conflict in our society between personal choice and governmental control? These are serious questions, for which today there are no satisfactory answers.

References

Avery, O. T., C. M. MacCleod, and M. McCarty. 1944. Studies on the chemical nature of the substance inducing transformation in pneumococcal types. *Journal of Experimental Medicine* 79: 137-158.

Becker, R. O. 1972. Stimulation of partial limb regeneration in mammals. *Nature* 235: 109-111.

Carlson, E. A. 1972. H. J. Muller, *Genetics* 70: 1-30.

Darwin, C. 1874. *The descent of man,* 2nd ed. D. Appleton Co., New York.

Fasten, N. 1935. *Principles of genetics and eugenics.* Ginn & Co., New York.

Friedmann, T. 1971. Prenatal diagnosis of genetic diseases. *Scientific American* 225 (5): 34-51.

Galton, F. 1908. *Inquiries into Human Faculty and Its Development,* 2nd ed. E. P. Dutton Co., New York.

_____. 1909. *Essays on eugenics.* Eugenics Education Society, London.

Gardner, E. J. 1972. *Principles of genetics,* 4th ed. John Wiley & Sons, Inc., New York.

Gurdon, J. B. 1968. Transplanted nuclei and cell differentiation. *Scientific American* 219 (6): 24-36.

_____. 1970. The transplantation of nuclei from single cultured cells into enucleated frog's eggs. *Journal of Embryological and Experimental Morphology* 24: 227-248.

Heim, W. 1972. Moral and legal decisions in reproductive and genetic engineering. *American Biology Teacher* 34 (6): 315-318.

McKusick, V. A. 1968. *Mendelian inheritance in man: catalogs of autosomal dominants, autosomal recessives and X-linked phenotypes,* 2nd ed. Johns Hopkins Press, Baltimore.

Merril, C. R., M. R. Geier, and J. C. Petricciani. 1971. Bacterial versus gene expression in human cells. *Nature* 233: 398-400.

Nadler, H. L., and A. Gerbie. 1971. Present status of amniocentesis in intrauterine diagnosis of genetic defects. *Obstetrics and Gynecology* 38: 789-799.

Steward, F. C., M. O. Mapes, and J. Smith. 1958. Growth and organized development of cultured cells, I: growth and division of freely suspended cells. *American Journal of Botany* 45: 693-703.

Taylor, G. R. 1968. *The biological time bomb*. World Publishing Co., New York.

Time [magazine]. 1971. Man into superman: the promise and perils of the new genetics. April 19:33-52.

Wallace, B., ed. 1972. *Essays in social biology, vol. 2*. Prentice-Hall, Inc., Englewood Cliffs, N.J.

8
Biochemical and Chromosomal Defects Yield to Prenatal Diagnosis

PAUL T. LIBASSI

Prenatal diagnosis of biochemical defects represents, according to Dr. John W. Littlefield, Chief of the Genetics Unit at Massachusetts General Hospital, an unusual situation in medicine. The diagnostic information which provides the basis for any therapeutic measures to be taken is based upon a single laboratory test for an enzyme deficiency. Because of this, laboratories must take inordinate precautions to develop successful cultures and enlist the aid of confirmatory procedures whenever possible. The preparation of successful cultures depends in great part on the amount of amniotic fluid from which cells are havested. Opinions on just how much fluid should be drawn vary.

Dr. M. Neil Macintyre's laboratory at Case Western Reserve University obtains as much as 40 cc of fluid, a quantity large enough to maintain cultures in as many as six or seven flasks. According to Dr. Macintyre, several important advantages result from multiple culture preparations. The possibility of an erroneous diagnosis due to maternal cell contamination is greatly minimized, and the likelihood of an earlier diagnosis is enhanced by the growth of a greater volume of cells.

Culture methods vary from laboratory to laboratory. A choice of culture techniques should ultimately depend upon those methods which supply the most accurate data in the shortest possible time, according to

Reprinted from *Laboratory Management.* Coypright United Business Publications, Inc. 1972.

Dr. Macintyre. The cytogenetic laboratory is responsible for production of cultures to be utilized for biochemical analysis and for chromosomal preparations of sufficient volume and quality to facilitate a speedy and accurate diagnostic evaluation. Dr. Littlefield's laboratory has found that a medium supplemented with 15 percent calf serum offers the best results. Cultures at his Massachusetts laboratory are re-fed twice a week, a procedure which appears to enhance the growth of even single cells.

Problems

Several types of cells from epithelial-like to fibroblast-like are present in amniotic fluid. Problems arise because clinicians cannot be certain that in the case of female fetuses the cells that have been cultured are fetal and not maternal. Researchers have suggested that several developments would alleviate this problem and confirm any discrepancies.

Specific and clear-cut differentiation of the isozyme content of fetal and adult cells would be of important practical value. Studies have documented differences in the activity of enzymes between cultured amniotic cells and adult skin fibroblasts. In addition, Dr. Littlefield suspects that different isozyme patterns, in addition to quantitative changes in enzyme activity, are displayed by amniotic cells. If an isozyme pattern peculiar to fetal cells could be defined, it would provide an important confirmatory adjunct. Dr. Littlefield adds that such an isozyme pattern should be detectable in a spread of relatively few cells so that confirmation could be made in time to permit drawing a second specimen of amniotic fluid if necessary.

An alternative to specific isozyme patterns to confirm fetal cells is visualization of the fetus before or during amniocentesis, thereby guaranteeing that fetal cells have in fact been obtained. Several scientists are molding fiber glass optics into an instrument called a "fetal amnioscope" just for such purposes.

Dr. Henry Nadler, Chairman of the Department of Pediatrics at Northwestern University and a pioneer in amniocentesis, reports that bioengineers at Northwestern have developed an instrument that can be inserted into a needle only slightly larger than the size presently used for amniocentesis. This instrument has facilitated visualization of an increased number of external morphologic anomalies and in the future may allow clinicians to actually remove a snip of fetal tissue assuring that cells being cultured are in fact fetal cells.

Refinements

In standard chromosomal analysis, the photomicrograph is enlarged and the chromosome images are cut out and arranged on a white card into 22 pairs of homologous chromosomes plus the two sex chromosomes. Pairs are arranged in sequential order according to morphological peculiarities based upon size, shape and ratio of arm length.

Several refinements of this manual procedure have since rendered the construction and examination of the karyotype both less time-consuming and more objective. Dr. Robert S. Ledley of the National Biomedical Research Foundation has developed a computer program which can analyze a photomicrograph with complete accuracy in less than 40 seconds.

The computer is programmed to count the total number of chromosomes in each cell and to measure and notate their morphological features. The computer then matches chromosome pairs according to these features and classifies each pair according to a standard karyotype sequence. In this manner, the computer is capable of evaluating small but clinically significant chromosomal abnormalities which are often undiscernible by the human eye. Dr. Peter W. Neurath of New England Medical Center, Boston, also has developed a computerized chromosome analysis system (*Lab. Mgt.,* Sept. 1970).

Dr. Ledley is presently developing a special-purpose computer that will automate the entire procedure from selection to analysis. Such a system would eliminate any degree of subjectivity involved in the selection of cells to be photographed for analysis. A scanner linked directly to the microscope would automatically select and photograph the appropriate cells. The only problem with the system at this point seems to be the costs involved. Dr. Harold Nitowsky of the Albert Einstein Medical Center explains, "In terms of cost analysis, the manual procedures are certainly more feasible at this point."

The discovery of banding patterns unique for each individual pair of human chromosomes has resulted in a reliable and accurate method for identifying and aligning chromosomes on a karyotype. Dr. Margery Shaw at the University of Texas has found that chromosomes stained with a Giemsa stain display discrete and highly specific banding patterns in characteristic sections of the chromosome arm. Not only does this technique permit accurate homologous pairing, but banding patterns can reveal chromosomal defects impossible to discern by morphological characterization alone.

Microassays

One of the largest problems of prenatal diagnosis involves the relatively large volume of cells required for biochemical assay and the time required to grow them. As long as six weeks may be needed to culture sufficient amounts of cells for the types of biochemical assays currently available. The development of microassays for the various enzymes involved would solve this problem to a great extent. Dr. Nadler is encouraged by the development of histochemical techniques sensitive enough to permit analysis of significantly fewer cells than are now required. If these techniques prove accurate, as few as five cells taking only three of four days to grow could be assayed.

Although analysis of uncultured amniotic cells is possible and has been performed with a degree of success, most work to date has been done with cultured cells. Researchers indicate that the reliability of a diagnosis on the basis of uncultured cells is questionable and should be avoided. Some cells harvested from the amniotic fluid are dead and can therefore lead to an erroneous diagnosis.

Uncultured cells are most frequently used to determine sex of the fetus as early as possible to facilitate the management of pregnancies in women heterozygous for X-linked disorders such as hemophilia. Amniotic fluid cells are harvested from cultures following the first day of incubation and are stained with fluorescent dyes for sex chromatin. This staining technique reveals the presence of a Y chromosome, indicating that the fetus is male. However, in cases where no fluorescence is present, conclusions based upon this technique alone are equivocal. The Y chromosome may have failed to fluoresce for a variety of reasons, the fetus may be female or only maternal cells may be present in the fluid. Furthermore, it is unlikely though possible that an X-linked disorder such as Turner's syndrome may interfer with a correct analysis of sex determination. For these reasons, researchers such as Dr. Macintyre recommend that sex chromatin analysis of uncultured cells be confirmed through karyotypes of cultured cells before diagnosis is finalized.

A problem area of particular concern to many clinicians results in cases of multiple pregnancies. In these situations there is a distinct possibility that amniotic fluid will be inadvertently sampled from only one of the amniotic sacs and the resultant cell cultures will therefore represent only one fetus. The development of ultrasonic techniques has recently made detection of multiple pregnancies possible early in gestation. As early as the eighth week of pregnancy ultrasonic probes can scan the surface of the abdomen and bounce back a series of echoes which can then be interpreted to identify and localize the fetuses.

Sound waves pass through abdominal tissue at variable speeds according to density. From the pattern of echoes bounced back and viewed on a television screen, doctors can determine the presence of multiple fetuses and pinpoint their positions in the uterus. Localization of the fetus affords information for a more effective amniocentesis and by alerting doctors to multiple pregnancies indicates the need to draw fluid from each sac and to maintain separate cultures.

Caution

Despite the fact that amniocentesis has come a long way in the last several years—two years ago an estimated 200-300 had been performed in the world; today some 1500 have been successfully carried out in the U.S. alone—most scientists and clinicians agree that the technique still remains rather crude and should not be undertaken by any except those with a reasonable amount of experience and expertise.

Dr. Nadler recommends that a series of requirements be satisfactorily fulfilled before intrauterine diagnostic measures are undertaken. These include an obstetrician experienced in the technique of amniocentesis, laboratory personnel experienced in cultivating amniotic fluid cells and in selecting and developing reliable spreads for karyotypes, technicians experienced in performing the specific diagnostic tests required, and the ability to provide suitable treatment. Treatment generally consists of therapeutic abortion. Dr. Henry Kirkman, Chief of Laboratories at North Carolina Memorial Hospital reports that amniocentesis is not performed on those patients for whom abortion is out of the question for either religious or personal reasons.

Diagnostic amniocenteses are recommended by physicians at the Genetics Referral Center of the University of California at San Diego on the basis of the following statistical portrait: a previous history of hereditary disorders; the mother is a known or suspected carrier of an X-linked disorder; the parents are known carriers of a chromosomal translocation; a previous child was born with a chromosome abnormality; the maternal age at conception was 35 or more years.

Not A Total Answer

Nor does amniocentesis provide the answers for all genetically related disorders. The two most common genetic defects, sickle cell anemia and cystic fibrosis, cannot as yet be detected prenatally with any degree of certainty. Methods for both of these are on the horizon and it seems only a matter of time before they are perfected.

Dr. Barbara Bowman at the University of Texas has demonstrated that serum from patients afflicted with cystic fibrosis inhibits the action of oyster cilia. The technique has not, however, yielded clear-cut separations between carriers and those afflicted with the disease. When these and other problems have been ironed out, Dr. Bowman hopes to adapt the technique to amniotic fluid for diagnosis of the disease *in utero.*

Prenatal diagnosis of sickle cell anemia presents different problems. According to Dr. Nitowsky, methods for identifying the sickle cell genotype are currently available, but prenatal diagnosis is dependent upon obtaining specimens of fetal blood, something which has never been attempted. With the development of the "fetal amnioscope" drawing of fetal blood may become a possibility.

If small amounts of fetal blood could be drawn, red cells from the fetus which normally manufacture Hemoglobin F could be "tricked" into synthesizing adult hemoglobin by inhibiting production of normal fetal hemoglobin. Once fetal hemoglobin is inhibited, the red cells would be compelled to produce Hemoglobin A and/or Hemoglobin S, indicating a propensity towards the disease. Using chromatography, cells could then be separated and an evaluation made.

9
Role of Amniocentesis in the Intrauterine Detection of Genetic Disorders

HENRY L. NADLER, M.D.,
AND ALBERT B. GERBIE, M.D.

Abstract. One hundred and sixty-two transabdominal amniocenteses were performed between the thirteenth and eighteenth weeks of fetal gestation as part of the management of 155 "high-risk" pregnancies. Successful cultivation of amniotic-fluid cells led to the intrauterine detection of Down's syndrome (10 cases), Pompe's disease (one case), lysosomal acid phosphatase deficiency (one case) and metachromatic leukodystrophy (one case). The risk of this procedure is low since neither fetal nor maternal complications were demonstrated in this series of patients. Cultivation of amnioticfluid cells obtained by transabdominal amniocentesis early in the second trimester of pregnancy provides a method that enables parents at "high risk" for having offsping with certain serious genetic disorders to have children without risk of such a defect.

Amniocentesis has been used as a diagnostic aid since the early 1930's.[1] Since the demonstration of its value in the management of Rh isoimmunization, the technic of transabdominal amniocentesis has gained widespread acceptance.[2] This procedure has been performed over 10,000 times after the twentieth week of pregnancy

Printed with permission from *The New England Journal of Medicine* Vol. 282, pp. 596-599, March 12, 1970.

and maternal or fetal morbidity or mortality reported in less than 1 per cent of cases.[3-7] In adverse effects that have been reported fetal mortality appears to be greater than maternal, with fetal deaths reported due to abruptio placentae, amnionitis and fetal hemorrhage.[5-7] Puncture of the fetus has been reported.[7-9] The maternal morbidity includes amnionitis, maternal hemorrhage, abdominal pain and peritonitis. More recently transabdominal amniocentesis has been performed early in the second trimester of pregnancy. In most cases, amniocentesis has been performed immediately before pregnancy was interrupted, making it difficult if not impossible to define the risks to either fetus or mother accurately.[10,11] Transvaginal amniocentesis has been shown to carry an appreciable risk of spontaneous abortion when performed early in pregnancy.[12]

During the past few years, sex-chromatin analysis and cultivation of amniotic-fluid cells obtained by transabdominal amniocentesis early in pregnancy have resulted in the detection of a number of genetic defects of the fetus, including X-linked recessive disorders,[12] Down's syndrome,[13,14] mucopolysaccharidosis,[13,15] Prompe's disease,[16] X-linked uric aciduria,[17] cystic fibrosis[18] and lysosomal acid phosphatase deficiency.[19]

The purpose of this study was to evaluate the usefulness of transabdominal amniocentesis and cultivation of amniotic-fluid cells in the management of pregnancies in which there is an appreciable risk that the child will be affected with a serious genetic disorder.

Methods and Materials

Transabdominal Amniocentesis

One hundred and sixty-two transabdominal amniocenteses were performed between the thirteenth and eighteenth weeks of fetal gestation in 142 patients during 155 pregnancies. Transabdominal amniocentesis was performed by an obstetrician as an outpatient procedure after thorough explanation of the risks and with signed permission of the pregnant woman and her husband. After the patient voids, her skin is cleansed with alcohol and benzalkonium (Zephiran) solution. Strict aseptic conditions are observed throughout the procedure. A local anesthetic, 1 per cent lidocaine (Xylocaine) solution, is injected into the proposed puncture site. A 22-gauge, 5-inch spinal needle with stylet is inserted through the abdominal wall in the midline, directed at a right angle toward the middle of the uterine cavity. After puncture, the stylet is removed, and a sterile plastic syringe is used to withdraw 10 ml of amniotic fluid, after which the needle is swiftly withdrawn.

Cultivation of Amniotic-Fluid Cells

The amniotic fluid is placed in either a sterile siliconized glass or a plastic tube and transported at ambient temperature to the laboratory. The fluid is centrifuged at 100 X g for 12 minutes, the supernatant removed and forzen, and the cell pellet suspended in 0.5 ml of 100 per cent fetal calf serum. The cells are placed in five small Falcon Petri dishes, immobilized under a glass cover slip, 2 to 3 ml of F-10 nutrient medium (BBL) supplemented with 15 per cent of fetal calf serum containing antibioticantimycotic mixture is added, and the cultures placed in 5 per cent carbon dioxide at 37°C. The medium is changed every other day until a number of colonies of cells are seen under the cover slips, usually seven to eighteen days. The medium is removed from the dish, and the cover slip turned cell surface up and placed in another Falcon dish. Two milliliters of medium is added to both the original dish and the new dish. The cultures may now be used directly for chromosome analysis or subcultured and used for chromosome or biochemical analysis.[15,16,19-22]

Chromosome Analysis

Twenty hours after the cover slip is inverted or after subculture onto a cover slip, 0.2 ml of diacetylmethyl colchicine (Colcemid), 0.01 mg per milliliter, is added. Four hours later the medium is removed, the cover slip rinsed gently with hypotonic solution (4 parts distilled water to 1 part medium) and then incubated in this hypotonic solution at 37° for 30 minutes. The cover slip is gently rinsed in a fixative (1 part glacial acetic acid to 3 parts absolute methanol) and placed in fresh fixative for 20 minutes. The cover slip is rinsed in 50 per cent acetic acid and placed cell side up on a slide. A few more drops of 50 per cent acetic acid are added to the cover slip, which is gently passed about 5 inches above a flame, permitting some of the acetic acid to evaporate. The excess acetic acid is drained into a blotter or tissue. The cover slip is placed in methanol for five minutes and stained for 10 minutes with Giemsa reagent.

Results

Transabdominal amniocentesis was successfully performed in 160 of 162 cases. Repeat amniocentesis was required in seven. Amniotic fluid was not obtained in the initial amniocentesis in two cases, and in the

TABLE 1.   Indications for Amniocentesis in 155 Pregnancies.

Indication	No. of Pregnancies	Outcome of Pregnancy
Chromosomal	132	
Translocation carrier	22	7 with Down's syndrome (therapeutic abortion); 15 normal.
Maternal age > 40 yr	82	2 with Down's syndrome (therapeutic abortion); 80 normal.
Previous trisomic Down's syndrome	28	1 with Down's syndrome; 27 normal.
Familial metabolic	23	
Carrier of X-linked recessive disorder	7	2 males (therapeutic abortion); 5 females (normal).
Pompe's disease	8	1 with Pompe's disease (therapeutic abortion); 6 normal. 1 normal (spontaneous abortion).
Lysosomal acid phosphatase deficiency	2	1 affected (therapeutic abortion).
Metachromatic leukodystrophy	1	1 affected (therapeutic abortion).
Mucopolysaccharidosis	2	2 normal
Generalized gangliosidosis	1	1 normal
Maple-syrup-urine disease	2	2 normal

remaining five, either contaminated amniotic fluid or inadequate cell growth necessitated a repeat amniocentesis.

Successful cultivation of amniotic-fluid cells was accomplished in 155 of 160 amniocenteses—that is, in all 155 pregnancies. The time between amniocentesis and successful chromosome analysis ranged from three to 28 days, with a mean of 14.2 days, as compared to a range of 15 to 40 days, with a mean of 30.1 days, for biochemical analysis. Successful cultivation of amniotic-fluid cells for more than three subcultures was accomplished in 75 per cent of all cases. In all but one the predicted sex was confirmed after delivery by examination of the external genitalia, and after therapeutic abortion by chromosome analysis of the cultured abortion material. In three cases when hypertonic saline was used for therapeutic abortion, chromosome analysis of the abortus was not possible, but repeat amniotic-fluid analysis demonstrated similar karyotypes. In one case the predicted sex of the baby was not confirmed at delivery. Chromosome analysis showing 46, XX, was completed four days after amniocentesis and confirmed at six days. The patient gave birth to normal male twins. Re-examination of the chromosome analysis of the amniotic-fluid cell culture, which had been frozen, after 45 days of growth, demonstrated a 46 XY karyotype. In this case and in another case prenatal chromosome analysis did not predict the presence of twins.

Table 1 lists the indication for the amniocentesis and the outcome of the pregnancy. In 10 patients Down's syndrome was detected in utero. Six pregnancies were terminated by hysterotomy, and three by intra-amniotic instillation of hypertonic saline. The diagnosis of Down's syndrome was confirmed in these nine abortions by examination or culture or both. In one case the parents elected to continue the pregnancy, and an infant with Down's syndrome was delivered at term. Two mothers who had previously borne children with X-linked muscular dystrophy were found to be carrying a male fetus. In each the pregnancy was terminated by intra-amniotic saline injection. In previously reported cases Pompe's disease[16] and lysosomal acid phosphatase deficiency[19] was detected in utero. The pregnancies were interrupted by hysterotomy, and the diagnosis confirmed by biochemical examination of the fetus. In one case aryl sulfatase A activity was deficient in cultivated amniotic-fluid cells obtained from a woman who had previously delivered a child with metachromatic leukodystrophy (65 nmoles of product released per hour per milligram of protein as compared to six controls with values of 825 to 1400 nmoles of product released per hour per milligram of protein). The pregnancy was terminated by hysterotomy, and aryl sulfatase A activity in the liver of the fetus was less than 5 per cent of normal. Enzyme activity in cultivated amniotic-fluid cells from the remaining pregnancies at risk for Pompe's disease, lysosomal acid phosphatase deficiency, generalized gangliosidosis and maple-syrup-urine disease was readily detectable, and normal children, presumably two thirds of whom may be carriers, were delivered. Cultivated amniotic-fluid cells obtained in two pregnancies from a patient at risk for Hurler's syndrome had no detectable metachromatic granules, and accumulation of mucopolysaccharide labeled with radioactive sulfate was within normal limits. The children are now six and 18 months of age, with no evidence of Hurler's syndrome either on physical or on laboratory examination.

No evidence of maternal morbidity was encountered. In one case, one month after transabdominal amniocentesis a spontaneous abortion occurred. Examination of the fetus and placenta revealed evidence of recent minimal chorioamnionitis (Dr. Kurt Benirschke). The cause of the abortion was listed as incompetent cervix. Seven children born to mothers past the age of 40 years were noted to have the following malformations: spina bifida, one; horizontal palmar crease, one; cleft soft palate, one; pilonidal dimple, three; congenital heart disease (patent ductus), one; and café-au- lait spots, one. In one mother who had previously given birth to a child with Down's syndrome a child with dislocated hips was delivered.

None of these congential malformations could have been detected with the methods used in this study. Seven infants were delivered between 34 and 37 weeks of pregnancy with birth weights below 4 lb, 10 oz (3 lb, 2 oz, 3 lb, 6 oz, 3 lb, 9 oz, 4 lb, 4 lb, 1 oz, 4 lb, 3 oz, and 4 lb, 6 oz).*

Discussion

Successful cultivation of amniotic-fluid cells for chromosome analysis was achieved in 97 per cent of cases with one amiocentesis—yielding 10 ml of amniotic fluid. Repeat amniocentesis resulted in successful cultivation in the five cases in which inadequate growth or contaminated specimen had not permitted successful chromosome analysis. Chromosome analysis was accurate in all but one case, in which, as well as in one reported by Uhlendorf,[23] the analysis was performed much earlier than usual because of the rapid cell growth. These cells, presumably maternal macrophages, die after approximately a week, and repeat examination of the cultures after two weeks will permit accurate chromosome analysis. In a case reported by Macintyre[23] one of five cultures from one patient had XX cells whereas the remaining cultures were XY. On the basis of these observations it is suggested that chromosome analysis be performed on at least two cultures at different times of cultivation. Another source of error is the inability to detect the presence of twins of similar or possibly even different sex on the basis of chromosome analysis.

Limited information is available concerning the reliability of intrauterine detection of biochemical disorders. In only one inborn error of metabolism, Pompe's disease, has a large enough number of pregnancies been studied to evaluate the reliability of this approach.[16] At present the major limiting factor appears to be the difficulty in obtaining an adequate number of cells for biochemical analysis.

The experience in this study suggests that transabdominal amniocentesis early in the second trimester of pregnancy carries minimal risks to mother and fetus. In addition, to the 155 patients in this study, 132 have been monitored in three other centers, without evidence of fetal or maternal complications.[23] Complications will undoubtedly occur but the risk will probably be about the same as that of transabdominal amniocentesis after 20 weeks of pregnancy,—that is, less than 1 per cent.

The ability to detect a genetic defect in the fetus not only gives new precision to genetic counseling but also creates many moral, legal and medical problems. At present, since effective treatment of genetic defects

*The summary of each patient is available upon request from the authors.

is limited, the question of interruption of pregnancy is raised. Interruption of pregnancy on the basis of fetal abnormality is legal in relatively few states, and in many countries it must be performed before 20 weeks of gestation. One medical problem that must be evaluated is the safety of thrapeutic abortion performed in the second trimester of pregnancy. The two procedures most often used for interruption of pregnancy in the second trimester are hysterotomy and intra-amniotic instillation of hypertonic solutions. The maternal morbidity and mortality of intra-amniotic hypertonic saline for therapeutic abortion include infection, intravascular injection of saline resulting in hypernatremia, post-partum hemorrhage, cervical lacerations, uterine rupture, cerebral infarction and sudden death after shocklike symptoms or vascular collapse.[2,4] However, the risk of this procedure can be kept to a minimum if careful attention is given to technical details. These include knowledge of the potential complications, a uterus of more than 15 weeks' gestational size and instillation of not more than 200 ml of 20 per cent saline.

The selection of patients for whom intrauterine monitoring for detection of a genetic disorder in the fetus is warranted will depend upon accurate assessment of the risk of the procedure and the reliability of diagnosis as compared to the risk of an affected fetus. The results of this study demonstrate the usefulness of intrauterine monitoring of a number of groups of "high-risk" pregnancies.

We are indebted to Sally Lee, Anita Messina, Cathy Ryan, Elvira Shannon and Marilyn Swae for technical assistance, to the many physicians who helped in obtaining this material, and to Drs. A. E. Emery, C. B. Jacobson, J. W. Littlefield, M. N. Macintyre and B. W. Uhlendorf for sharing their experience with us and for suggestions.

References

1. Menees TO, Miller JD, Holly LE: Amniography: preliminary report. Amer J Roentgen 24:363-366, 1930
2. Liley AW: The use of amniocentesis and fetal transfusion in erythroblastosis fetalis. Pediatrics 35:836-847, 1965
3. Freda VJ: Recent obstetrical advances in the Rh problem: antepartum management, amniocentesis, and experience with hysterotomy and surgery in utero. Bull NY Acad Med 42:474-503, 1966
4. Queenan JT: Amniocentesis and transamniotic fetal transfusion for Rh disease. Clin Obstet Gynec 9:491-507, 1966
5. Burnett RG, Anderson WR: The hazards of amniocentesis. J Iowa Med Soc 58:130-137, 1968

6. Liley AW: The technique and complications of amniocentesis. New Zealand Med J 59:581-586, 1960

7. Creasman WT, Lawrence RA, Thiede HA: Fetal complications of amniocentesis. JAMA 204:949-952, 1968

8. Berner HW Jr: Amniography, an accurate way to localize the placenta: a comparison with soft-tissue placentography. Obstet Gynec 29:200-206, 1967

9. Wiltchik SG, Schwarz RH, Emich JP Jr: Amniography for placental localization. Obstet Gynec 28:641-645, 1966

10. Jacobson CB, Barter RH: Intrauterine diagnosis and management of genetic defects. Amer J. Obstet Gynec 99:796-807, 1967

11. Nadler HL: Medical progress: prenatal detection of genetic defects. J. Pediat 72:132-143, 1969

12. Fuchs F: Genetic information from amniotic fluid constituents. Clin Obstet Gynec 9:565-573, 1966

13. Nadler HL: Antenatal detection of hereditary disorders. Pediatrics 42:912-918, 1968

14. Valenti C, Schutta EJ, Kehaty T: Prenatal diagnosis of Down's syndrome. Lancet 2:220, 1968

15. Fratantoni JC, Neufeld EF, Uhlendorf BW, et al: Intrauterine diagnosis of the Hurler and Hunter syndromes. New Eng J Med 280:686-688, 1969

16. Nadler HL, Messina AM: In-utero detection of Type-II glycogenosis (Pompe's disease). Lancet 2:1277-1278, 1969

17. DeMars R, Sarto G. Felix JS, et al: Lesch-Nyhan mutation: prenatal detection with amniotic fluid cells. Science 164:1303-1305, 1969

18. Nadler HL, Swae MA, Wodnicki JM, et al: Cultivated amniotic-fluid cells and fibroblasts derived from families with cystic fibrosis. Lancet 2:84-85, 1969

19. Nadler HL, Egan TJ: Deficiency of lysosomal acid phosphatase: a new familial metabolic disorder. New Eng J Med 282:302-307, 1970

20. Porter MT, Fluharty AL, Kihara H: Metachromatic leukodystrophy: arylsulfatase-A deficiency in skin fibroblast cultures. Proc Nat Acad Sci 62:887-891, 1969

21. Yarborough DJ, Meyer OT, Dannenberg AM Jr, et al: Histochemistry of macrophage hydrolases. 3. Studies on beta-galactosidase, beta-glucuronidase and aminopeptidase with indolyl and naphthyl substrates. J Reticuloendothel Soc 4:390-408, 1967

22. Dancis J, Hutzler J, Cox RP: Enzyme defect in skin fibroblasts in intermittent branched-chained ketonuria and in maple syrup urine disease. Biochem Med 2:407-411, 1969

23. Emery AEH, Jacobson CB, Macintyre MN: Personal communication

24. Schiffer MA: Induction of labor by intra-amniotic instillation of hypertonic solution for therapeutic abortion or intrauterine death. Obstet Gynec 33:729-736, 1969

10
Human Evolution by Voluntary Choice of Germ Plasm

H. J. MULLER

For some decades the term *eugenics* has been in such disrepute, as a result of its spurious use in support of the atrocities committed by those with class and race prejudices, that few responsible students of evolution or genetics have dared to contaminate themselves by mentioning it, much less by dealing with the subject except in condemnation. However, it is now high time to take new stock of the situation. For the odious perversions of the subject should not blind us longer to a set of hard truths, and of genuine ethical values concerning human evolution, that cannot be permanently ignored or denied without ultimate disaster. On the other hand, if these truths are duly recognized and given expression in suitable policies, they may open the way to an immeasurable extension and enhancement of the potentialities of human existence.

In view of the signal defeat in World War II of the leading exponents of racism—a defeat which is still gathering momentum—and the declining prestige afforded in the Western world to the claims of aristocratic or bourgeois class differentiations, it at last becomes feasible to return, in a more reasonable spirit, to the theme of prospective human biological evolution. Moreover, for this job of reexamination we are now provided not only with a better understanding of genetic and evolutionary principles but also with a considerably reformed structure in most Western societies, liberalized mores, a heightened freedom of discussion, and a marked improvement in technologies, all of which combine to make possible approaches that earlier would have seemed out of the question.

Science, 1961, Vol. 134, Issue 3480, pp. 643-649. Copyright © 1961 by the American Association for the Advancement of Science.

It was Darwin who pointed out that modern culture is causing a relaxation and perhaps even a reversal of selection for socially desirable traits, and he expressed himself rather pessimistically about the matter, although in his time this process must have been much less pronounced than it is nowadays. His cousin Galton, impressed by Darwin's arguments concerning evolution in general as well as by those pertaining to man, but unwilling to accept defeat or frustration for humanity on this score, proposed the idea that the trend might be counteracted consciously. For this course of action he coined the term *eugenics,* included within which he understood all measures calculated to affect the hereditary constitution in a favorable way. As he pointed out, these measures might be of very diverse kinds, lying not only in such fields as medicine but also in education, economics, public policy in general, and social customs, although he did not comtemplate drastic changes from the mores of that Victorian age.

Unfortunately, although Galton realized to some extent the influence of the social and familial environment in the shaping of people's psychological traits, he was not sufficiently aware of the profundity of the environmental control. He therefore made the naive mistake, so widespread in his day, of looking upon the performances of different ethnic, national and social groups as indicative of their genetic capabilities and inclinations, although there were plenty of object lessons of the comparatively rapid transference of cultures that should have taught him better. Later, it was the madness of such out-and-out racists and so-called "social Darwinists" as Madison Grant, Lothrop Stoddard, Eugen Fischer, Lenz (1), and the Hitlerites which, carrying these prejudices much further, brought such odium upon the whole concept of eugenics as to run it into the ground.

Meanwhile, a large group of psychologists, represented by the Watson school, and of other social scientists, social reformers, socialists, and communists, all of them persons of egalitarian sympathies, impressed by the enormous potency of educational and other cultural influences, and regarding all eugenics as a dangerous kind of reaction that threatened their own roads to progress, popularized the idea that differences in human faculties are of negligible consequence not only as between different peoples and social classes but even as between individuals of the same group. They held that genetics in man could be allowed to take care of itself. And even where some genetic defects were admitted to exist, it was maintained that improved medical, psychological, and other cultural ministrations would provide sufficient remedies for them. Moreover, added the many Lamarckians among these groups, the improvements thereby acquired would eventually pass into the hereditary constitution. In this way, not only would all men become equalized but they would rise to ever higher biological as well as cultural levels. By about 1936 it had become a dire heresy among the official communists to dispute this line of argument, and the word *eugenics* had become a favorite symbol of all that is vile.

It is no wonder that earnest students of evolution and genetics, confronted by the mighty currents of these contending movements, which had advanced into the area of power politics, seeing how intertwined truth and error had become, and aware that their own views would almost certainly be misconstrued, tended to withdraw into their ivory towers and to refuse to discuss seriously the possible applications of genetics to man. It is to their credit, however, that only a few floated with the current that happened to be around them. But even fewer tried to contend with that current by raising the voice of reason, for that way lay the path to martyrdom.

Perhaps the last attempt made, up until the past few months, to present an appraisal of eugenics undistorted by extremist politics was the drafting of the "Geneticists' Manifesto" (2) of 1939, signed by about a score of participants at the International Genetics Congress held in Edinburgh just as the curtain began to rise on World War II. In this document it was pointed out that by far the greatest causes of differences between human groups in regard to psychological traits were environmental, predominantly cultural, whereas in the causation of such differences between individuals within the same group, both environmental and genetic factors were very powerful, and often comparable in their potency of action. The need for far-reaching reforms—for affording more nearly equal opportunities to all groups as well as to all individuals and for removing biases—was stressed in this document, not only for the sake of persons directly concerned but also to provide a groundwork for the truer assessment of genetic differences, in the interest of more soundly based eugenics.

These reforms in society were also needed, it was pointed out, for the attainment by the population of a sounder set of values, applicable equally to eugenic and to cultural purposes: values by which active service and creativity would be regarded more highly than either passive submissiveness or self-aggrandizement. The "Geneticists' Manifesto," far from discarding the concept of eugenics per se, acknowledged that it afforded, when rightly used, a means of making far-reaching human progress of a kind that must complement purely cultural advancement. Beyond that, genetic improvement was even affirmed to be a right which future generations would consider those of the past who were aware of the situation as having been obligated to accord them, just as they in turn would consider themselves as being similarly obligated to their own successors. This obligation would not be regarded as a burden, however, but rather as a high privilege and a challenge to their creativity. At the same time, it was recognized that adequate implementation of eugenic policies also required a clearing away of the ancient heritage of superstition and taboos that hitherto had so obstinately enshackled human usages and preconceptions in matters of sex and reproduction.

It is true that our own world of today is still grievously beset with the old inequalities, prejudices, and mummeries. However, all peoples have by now seen the handwriting on the wall that spells the end of these irrationalities. For modern technologies have, on the one hand, made it too dangerous for the world to remain divided. They have, on the other hand, provided the means for achieving an unparalleled interdiffusion of techniques, ideas, personnel, education, and socioeconomic organization, and for raising standards of living. In the process, provincialisms are at last being ground down, though not without much friction. Opportunities, educational, economic, and social, are being extended ever more effectively to the more depressed social classes and ethnic groups in our own country and elsewhere. A real effort is being made to bring the viewpoint of science home to the general population. The battles that superstition is still winning take place ever closer and closer to its heartland, as was so well depicted in the moving picture on the Scopes trial, *Inherit the Wind.* And it is even becoming permissible to debate seriously matters that in the days before nuclear weapons, space ships, Kinsey, and the Darwin centenary were taboo among all nice people.

Thus the scene has at last been shifted to such an extent as to make it fitting to re-examine even such a scandalous subject as eugenics, with a view to preparing the new forces now arising in the world to deal with it both realistically and humanistically. For, as we shall see, cultural progress of the kinds mentioned has already proceeded far enough here and there to make the beginnings of a new approach to the subject possible, and the attitudes now formed and the preparations now made may presently lead, when the time is riper, to more salutary developements in this field than could ever before have occurred

Contradictions in the
Traditional Eugenic Methods

Let us first examine the methods by which it has hitherto been thought that eugenics might operate. These methods have taken their cue from the natural selection of the past. All evolution has had its direction determined in some way by the force of selection. Selection chooses among the materials available to it, namely, diverse mutations, which occur in a manner that is fortuitous so far as their adaptation to the needs of the organism in the given situation is concerned. Selection acts entirely through differential multiplication, but this process can be conceptually divided into two parts, namely, unequal rates of survival (or, conversely stated, of mortality), on the one hand, and unequal rates of reproduction of the survivors, that is, differential fertility (or, conversely stated, differential infertility), on the other hand.

Eugenists have therefore distinguished between two conceivable methods—differential control over mortality, and differential control over reproductive rate. The first method, however, although practiced by the Spartans and by primitive tribes who destroyed infants regarded as undesirable, is universally acknowledged to be inconsistent with the respect for human beings that forms an essential part of civilization. It might be contended that artificial abortion is an intermediate method, but everyone recognizes this also to be an undesirable means where other procedures are available. Essentially, then, this has left for eugenics the second alternative: that of a qualitatively differential control over reproduction prior to or at conception.

In Galton's time, before the advent of modern contraceptive techniques, it was indeed rather visionary to conceive of people's reproduction being governed in the interests of the progeny. For this could be done only by the drastic method of surgical sterilization, of a type that interfered with the sexual life, or by such consummate self-control as voluntary abstention from intercourse, or from its completion.

The development of techniques for cutting or ligating the tubes that conduct the mature reproductive cells afforded less objectionable means of sterilization, but this procedure was still usually regarded by most people—rightly or wrongly—as too irrevocable, except, perhaps, for persons who were mentally or morally hopelessly irresponsible. For them, enforced operations of this type were legalized in some regions, although it was rightly pointed out that there was grave danger of abuse of the practice unless it were confined to the most extreme cases. For attitudes that seem wrong in one place or setting may seem right elsewhere, and nonconformists may at times have moral standards superior, in a longer perspective, to those of the majority who condemn them. Thus, the amount of sterilization resulting from legal applications of an advisable kind would be so minute as to have very little eugenic influence.

However, the invention of fairly practicable artificial means of voluntary contraception opened up much wider possibilities for the control of reproduction in economically developed countries. As we all know, advantage has been taken of these techniques on a large scale, and they have become one of the indispensable procedures whereby the general standard of living has been so greatly raised. Still more practicable means of contraception seem at last to be on the way, thanks to the efforts of a handful of devoted scientists, and they cannot come too soon, for it is imperative to make similar benefits possible in the less developed regions.

But although contraception that is used for the enhancement of cultural benefits through the control of population quantity is at the same time a *potential* instrument for the improvement of genetic quality, such improvement does not occur unless the contraception is specifically aimed in this direction. To be sure, this purpose might be achieved if the individual couples concerned were to reach their decisions about how many children to have in a highly idealistic spirit, one guided by almost heroic self-criticism and wisdom. We shall presently consider whether or not it is realistic to expect this. A second proposal has been that of altering the economic and social system in such a way that pople of higher gifts and greater natural warmth of fellow feeling—that is, the genetically more highly endowed—would be normally led into occupations and modes of life more conducive to having a large family. Conversely, the organization of society which this view would hold to be ideal would tend to lead persons less well endowed to choose, of their own accord, situations in life that would encourage them to expend their energies in other pursuits than reproduction, and would give them less inducement for raising families.

These two approaches, the individual and the societal are, of course, not mutually exlusive, and most 20th-century eugenists have advocated a combination of the two. But let us examine each of them more closely. First, as regards the individual approach, which is supposedly to be adopted by people in general once they have been well educated in matters of evolution and genetics, it should be acknowledged that poeple in general can in fact be taught to take pride in making great sacrifices for what they recognize to be a great cause, especially when they win social approval thereby. This has often happened in times of war as well as after social revolution. However, it seems asking almost too much to expect those individuals who are *really* less well equipped than the average, in mentality or disposition, to acknowledge to themselves that they are genetically inferior to their neighbors in these respects, and then to publicly admit this low appraisal of themselves by raising no family at all or a smaller one than normal, especially since at the same time they would often be thwarting a natural urge to achieve the deep fulfillments, accorded to their neighbors, that go with having little ones to care for and bring up. Moreover, those with physical impairments would likewise tend to rationalize the situation, by thinking that they possessed some superior psychological qualities that more than compensated for their physical defects.

In fact, then, the ones most likely to comply with the idea of restraining their own reproduction would be those who had such strong social feelings, such a sense of duty, so high a standard of what is good, so little egotism, and such an urge for objectivity, as actually to lean over backward and so underrate themselves. Thus we would be likely to lose for the next generation much of what might have been its best material.

On the other hand, for many of the really gifted there are often unusual opportunities for achievement, for rich experiences, and for service along other lines than those of bringing up a large family. Hence it would be only human of them, even though they were in sympathy with eugenic ideals, to expend a larger share of their energies in these other ways than does the average man or woman, for whom the home is often both a refuge and the chief stage on which to express leadership. In view of these considerations, it is not at all surprising that eugenic practices of this intentional, personal type, that require a correlation between the size of one's family and one's realistically made appraisal of one's genetic endowments, have made so little headway, even where they were approved theoretically. Thus, even among eugenists themselves, one seldom finds much evidence that these principles are being acted upon.

What, then, about the proposal that our society should introduce features into its structure whereby the more gifted, the abler, and the more socially minded would find conditions more conducive to their raising a large family, while those less capable or relatively antisocial would tend automatically to be deflected from family life? Surely we would not want a dictatorship to institute such a system, for dictators are oftener wrong than right in their decisions. Moreover, their subjects are not able to become truly men, in the all-around sense, and those who shine under such circumstances are not likely to be the wise and the responsible.

Under a democracy, on the other hand, is it not likely that "the common man" will refuse to subject himself to such manipulations? Certainly if the proposal took some such crude form as a subsidy for the raising of children, alloted to those who already occupied better positions or who had scored higher on certain tests, it would rightly be resented and defeated as dicriminatory by the great majority. And even if subtler forms of influence were used, such as special aids to family life for those in occupations requiring greater skills, responsibility, or sacrifice, there would soon be a clamor on all sides to have these advantages extended to every responsible citizen. No, we can hardly use democracy to support any kind of aristocracy. To be sure, ways might eventually be found to reduce the present strong negative correlation between educational or social achievement and size of family, and these would be all to the good. But no major formula is in sight for restoring the greater family size of the fitter, while retaining that most essential feature of our culture—the extension to all of mutual aid based on the most advanced technologies available.

It might seem to follow that we have now, as a result of our improved techniques for living, reached an inescapable genetic cul-de-sac. It might be concluded that we should therefore confine ourselves entirely to the immediate job on hand—the pressing and rewarding one of all social reformers and educators—that of making the best of human nature as it is, the while allowing it to slide genetically downhill, at an almost imperceptible pace in terms of our mortal time scale, hoping trustfully for some miracle in the future.

The New Approach: Germ-Cell Choice

However, it is man who has made the greatest miracles of any species, and he has overcome difficulties arising from his technologies by means of still better basic science, issuing in still better technologies by means of case of the genetic cul-de-sac of the present day, he has even now possessed himself of the means of breaking through it. For he is no longer limited, like species of the past which had the family system, to the two original methods of genetic selection applying to them: that of differential death rate on the one hand, and differential birth rate or family size, on the other hand. He has now given himself, in addition, the possibility of exerting conscious selection by making his own choice of the source of the germ cells from which the children of his family are to be derived (3–5). At present, this choice is confined to the male germ cells, but there are indications that with a comparatively small amount of research it might in some degree be extended to those of the female as well.

It is pretty common knowledge nowadays that some tens of thousands of babies have already been born, in the United States alone, that were derived by "AID," that is, by artificial insemination with sperm obtained by the physician from a donor chosen by him, but whose identity was kept unknown to all others, including even the parents. In the great majority of these cases the husband had been found to be irremediably sterile. And although, in view of the prevalence of the traditional mores, the whole matter was kept secret from the children themselves, both members of the couple had in these cases been eager to avail themselves of this opportunity to have one or more children. This method has, of course, been allowed only to those likely to make good parents—or, shall we say, good "love parents," as distinguished from "gene parents"? Moreover, follow-up studies have shown that these parents did truly love their children, as the children did their parents. It is noteworthy that this proved to be as much the case for the father as for the mother, and that the marriage was strengthened thereby.

Here, then, we see repeated what is typically found in those cases of early adoption in which the children have been genuinely desired. However, this "pre-adoption" (as Julian Huxley has termed it) is likely to prove even more binding and satisfying than "post adoption." And the method of pre-adoptional choice, despite its relative crudity at the present time, has demonstrated its capability of producing a superior lot of children. Thus, the couples that practice it have made a virture of necessity by inducing the genesis of children of whom they can usually be even prouder than of the children they would have had if they had been free from reproductive infirmity.

Recently, as more people have become alive to matters of genetics, an increasing number of couples have resorted to AID when the husband, although not sterile, was afflicted with some probably genetic impairment, or likely to carry such an impairment that had been found in his immediate family. Similarly, those with incompatabilities in blood antigens have also made use of the method with good results. In these ways a beginning has already been made in the conscious selection of germinal material for the benefit of the progeny. It is to be expected that many more people will seek such advantages for their prospective children when there are available for creating them the germ cells of persons who are decidedly superior in endowment to the normally healthy and capable persons that are commonly sought as donors by the physicians of today. For there is no physical, legal, or moral reason why the sources of the germ cells used should not represent the germinal capital of the most truly outstanding and eminently worthy personalities known, those who have demonstrated exceptional endowments of the very types most highly regarded by the couple concerned, and whose relatives also have tended to show these traits to a higher-than-average degree. How happy and proud many couples would be to have in their own family, to love and bring up as their own, children with such built-in promise.

Today, of course, most people have such an egotistical, individualistic feeling of special proprietorship and prerogative attaching to the thought of their own genetic material as to be offended at the suggestion that they might engage in such a procedure. No one proposes that they do so as long as they feel this way about the matter. However, they should not try to prevent others who would welcome such an opportunity from ordering their lives in accord with their own ideals. And as the prejudice against the practice gradually dwindles, the manifest value of the results for those who had participated in it would appeal to an increasingly large portion of the population.

In this connection it is important to bear in mind that there is no such thing as a paternal instinct in the sense of an inherent pride in one's own genetic material or stirps. Some primitive peoples, even including a few still in existence in widely separate regions, have had, strange as it may seem to us, no knowledge that the male plays a role in the production of the child, much less have they had any conception of genetic material or genes. Thus, among some of them the mother's brother has effectively filled the role of father in regularly caring for mother and children. Moreover, among some peoples, such as the Hawaiians, children are rather freely adopted at an early age into other families, into whose bosom they are warmly, fully, and unambiguously accepted as equals in every way to the natural children. It is "second nature," but not "first nature," for us in our society to exalt our own stirps.

It is, however, "first nature" for men and women to be fond of children and to want to care for them, and more especially, those children with whom they have become closely associated and who are dependent on them. If the love of a man for his dog, and vice versa, can go to such happy lengths as it often does, how much stronger does the bond normally become between the older and younger generations of human beings who live together. And since, in the past, the children *have* usually been those of the parents' own stirps, it has been a natural mistake to suppose that these stirps, rather than the human associations of daily life, formed the chief basis of the psychological bonds that existed between parent and child. Yet, as our illustrations have shown, this view is incorrect, and a family life of deep fulfillment can just as well develop where it is realized that the genetic connection lies only in our common humanity.

The wider adoption of the method of having children of chosen genetic material rather than of the genetic material fate has chanced to confer on the parents themselves implies, of course, that material from outstanding sources become available by having it stored in suitable banks (*3, 6, 7*). It would be preferable to have it in that glycerinized, deep-frozen condition developed by modern technology, in which it remains unchanged for an unlimited period without deterioration. It is true that research is badly needed for finding methods by which immature germinal tissue can, after the deep-freezing which it is known to survive, be restored to a state where it will multiply in vitro and subsequently produce an unlimited number of mature reproductive cells. Even without this further development, however, the way is already open, so far as purely biological considerations are concerned, for gathering and inexpensively storing copious reserves of that most precious of all treasures: the germinal material that has formed the biological basis of those human values that we hold in highest regard.

The high potential service to humanity represented by the pre-adoption of children should carry with it the privilege, for the parents, of having a major voice in choosing from what source their adopted material is to be derived. Surely, if they have ever had the right to produce, willy-nilly, the children that would fall to their lot as a result of natural circumstances, they should have the right of choice where they elect to depart from that haphazard method. They certainly would not wish knowingly to propagate manifest defectives, and, being idealistic enough to undertake this service at all, they would in most cases be glad to give serious consideration to the best available assessments of the genetic probabilities involved, as well as be open to advice regarding relative values and needs.

It would be made clear to parents that there is always an enormous amount of uncertainty concerning the outcome in the genetics of an organism so crossbreeding as man, especially since the most important traits of man are so greatly influenced by his cultural environment. Nevertheless, facing this, they would realize that the degree of promise was in any such case far greater than for those who followed the traditional course. It would be in full awareness of this situation that they would exercise their privilege of casting the *loaded* dice of their own choosing.

This kind of choice means that the physician can no longer be the sole arbiter of destiny in this matter. Clearly, if the couple are to accept their share of the responsibility and privilege here involved, the practice of keeping the donor unknown to them must be relinquished in these cases. Knowledge of the child's genetic lineage will also be needed later, so that sounder judgments may be reached concerning his genetic potentialities in the production of the generation to follow his own. This lifting of the veil of secrecy will become ever more practicable, and in fact even necessary, as the having of children by chosen genetic material becomes more widely accepted and therefore more frequent. Moreover, the attitude of others toward the couples who have employed this means of having their children will gradually become one of increasing acceptance and then of approbation and even honor.

Today's fear that knowledge by the mother of the identity of the gene father may lead to personal involvement between the two, to the detriment of normal family life, will recede when the gene source is remote in space or time, as when the germinal material has been kept in the deep-frozen state for decades. This procedure will also allow both the individual worth of those being considered as donors, and their latent genetic potentials, to be viewed in better perspective, and will reduce the danger that choices will be based on hasty judgments, swayed by the fads and fashions of the moment.

Let us see in what ways this method of reproduction from chosen material tends to avoid the difficulties that are encountered in attempting to reconcile traditional reproduction with the interests of genetic quality by somehow controlling the size of families. For one thing, as previously pointed out, most men would resist accepting and acting on the conclusion that they are below their next-door neighbor, or below average, in genetic quality. They would particularly resist the idea that they themselves are in that lowest fifth which would be required to refrain from having children if an equilibrium of genetic quality were to be maintained in the face of a 20-percent mutation rate. Yet most of these same people would willingly accept without resentment the idea that they are not among the truly exceptional who conform most closely to their own ideals. And so, when encouraged by the community mores, they would be glad and proud to have at least one of their children derived, by choices of their own, from among such sources. Thus they would continue to have families of a size more nearly conforming with their inclinations.

On the other hand, the worthy but humble, those who might otherwise, from overconscientiousness, limit their families unduly, would often be eager to serve as love parents. And although in that capacity they would tend to derive the germ cells from outside sources, they would be especially likely to have a well developed sense of values and so to choose sources even worthier than themselves. Finally, the really highly endowed but realistic would not be confronted with the sore dilemma of choosing between exercising their special gifts, on the one hand, or having the large family to which genetic duty seemed to obligate them, on the other hand. For their germinal material would tend to be sought by others, if not in their own generation, then later, and to a degree more or less in proportion to their achievements. Thus they would be freed to give their best services in whatever directions they elected.

In all these ways, the diverse obstacles encountered by a eugenics that tried to function by means of a consciously differential birth rate—that is, by adjustment of family size—would be avoided, Thereby, a salutary separation would be effected between three functions that often have conflicting needs today. These are, first, the choice of a conjugal partner; this should be determined primarily by sexual love, companionability, and compatible mentality and interests. Second, there is the determination of the size of the family; this should depend largely on the degree of parental love that the partners have, and on how successfully they can express it. Third, there is the promotion of genetic quality, both in general and in given particulars; these qualities are often very little connected with the first two kinds of specifications. By thus freeing these three major functions from each other, all of them can be far better fulfilled. Under these circumstances the conjugal partners need not be chosen by criteria in which a compromise is sought between the natural feelings and considerations of eugenics. Neither need the family size be restricted or expanded according to eugenic forebodings or feelings of duty. Yet at the same time there can be a far more effective differential multiplication of worthy genetic material than in any other humanly feasible way.

Further Prospects

It is likely that the avoidance of the effects of sterility will not be the only door through which such a change of mores will be approached. Facilities for keeping germ cells, suitably stored below ground in a deep-frozen condition, in areas relatively free from radiation and chemical mutagens, may well be provided in our generation for an increasing number of people (8). Among those would be persons subject to the growing radiation hazards of industry, commerce, war, and space flight, and those exposed to the as yet unassessed hazards of the chemical mutagens of modern life. The same means would greatly retard the accumulation of spontaneous mutations which probably occurs during ordinary aging. Thus, wives may in time demand such facilities for the storage of their husbands' sperm. These facilities would be provided not only for the sake of reducing mutational damage but also as a kind of insurance in the event of the husband's death or sterility. In these ways great banks of germinal material would eventually become available. They would be increasingly used not only as originally intended but also for purposes of conscious choice. Moreover, some of these stocks might become recognized as especially worthy only after those who had supplied them had passed away.

The cost of storage is, relatively, so small that failure to make such a provision will eventually be considered gross negligence. As Calvin Kline (7) points out, this will be especially the case where (as in India today) voluntary vasectomy becomes more prevalent as the surest and, in the end, the cheapest means of birth control. For when vasectomy is complemented by stores of sperm kept in vitro, the process of procreation thereby achieves its highest degree of control—control not subject to the impulses of the moment but only to more considered decisions.

It is true that most people's values, in any existing society, are not yet well enough developed for them to be trusted to make wise decisions of the kind needed for raising themselves by their bootstraps, as it were (9). But this type of genetic therapy, of "eutelegenesis," as Brewer termed it when he advocated it in 1935 (4), is certainly not going to spring into existence full fledged overnight. It will first be taken up by tiny groups of the most idealistic, humanistic, and at the same time realistic persons, who will tend to have especially well developed values. This mode of origination of such practices was pointed out by Weinstein in "Palamedes" in 1932 (10). These groups will tend to emphasize the most basic values that are distinctive of man, those that have raised him so far already, but which still may be enormously enhanced. Foremost among these are depth and scope of intelligence, curiosity, genuineness and warmth of fellow feeling, the feeling of oneness with others, joy in life and in achievement, keenness of appreciation, facility in expression, and creativity.

Those who follow these lodestars will blaze the trail, and others will follow and widen this trail as the results achieved provide the test of the correctness with which its direction was chosen. Meanwhile, the world in general, through its reorganizations of society and education, is moving in the same direction, by a de-emphasis of its provincialisms and a consequent recognition of the supreme worth of these basic human values. For after all, these values have been prominent in the major ethical systems of the whole world.

At the same time, plenty of diversity will inevitably be developed. For each especially interested group will naturally seek to enhance its particular proclivities, and this is all to the good. But on the whole, the major gifts of man have been found to be not antagonistic but correlated. Thus we may look forward to their eventual union with one another in a higher synthesis. And from each such synthesis in turn, divergent branches will always be budding out, to merge once more on ever higher levels.

It may be objected that we have next to no knowledge of the genes for those traits we value most, and that their effects are inextricably interwoven with those of environment. As was acknowledged earlier, this is quite true. However, it has also been true in all the natural selection of the past and in the great bulk of artificial selection. Yet these empirical procedures, based entirely on the accomplishment of the individuals concerned, did work amazingly well. We can do a good deal better by also taking advantage of the evidence from relatives and progeny. Yet that evidence also is furnished mainly by accomplishments or output. Where those were high, the environment was, to be sure, usually favorable, but so was the heredity. And as, in our human culture, social reform proceeds and opportunities become better distributed, our genetic judgments will become ever less obscured by environmental biases, while at the same time our knowledge of genes will improve. Meanwhile, the efforts of educators and the lessons of world affairs will serve to emphasize the same values for us. And these attitudes we will take over for our genetic judgments also. In this connection, another consideration deserves mention here. The preference which most parents will inevitably have for the genes of persons of truly remarkable achievement and character, rather than for those of the merely eminent or powerful, will at the same time serve to direct the stream of genetic progress toward the factors underlying creativity, initiative, originality, and independence of thought, on the one hand, and toward genuineness of human relations and affections, on the other hand. Otherwise the genetic movement might, as so often happens in other affairs of men, become directed toward skill at conformity, showmanship, and the dignified hypocrisy that often brings mundane success and high position (*9, 10*). This would have been a far greater danger in the case of the old-style eugenics.

But, it may be objected, does all this really represent conscious control in an over-all sense? Is it not merely a type of floating along in a chaotic manner, each straw making its own little movement independently of the rest, without a general plan or goal or stream? The answer is that humanity is as yet too limited in knowledge and imagination, too undeveloped in values, to see more than about one step ahead at a time. That step, however, can be discerned clearly enough, and by enough people, to give rise to a general trend in a salutary direction. And at the higher level to which each step taken will bring us we will be able to see an increasing measure of advance ahead. So we humans will achieve, not through dictation but through better general understanding and ever more clearly seen values, increasing mutual consent both concerning the means to be used and the aims toward which to orient. Thus an ever wider over-all view will emerge, and a surer, greater over-all plan, or rather, series of plans. To create them and to put them into effect will then enlist our willing efforts. And the very enjoyment of their fruits will bring us further forward in our great common endeavor: that of consciously controlling human evolution in the deeper interests of man himself (11).

References and Notes

1. H. J. Muller, *Birth Control Rev.* **17**, 19 (1933).
2. *J. Heredity* **30**, 371 (1940); *Nature* **144**, 521 (1939).
3. H. J. Muller, *Out of the Night* (Vanguard, New York, 1935).
4. H. Brewer, *Eugenics Rev.* **27**, 121 (1935); *Lancet* **1**, 265 (1939).
5. H. J. Muller, *Perspectives in Biol. and Med.* **3**, 1 (1959); ●●●●●●, in *Evolution after Darwin*, Sol Tax, Ed. (Univ. of Chicago Press, Chicago, 1960), vol. 2, p. 423.
6. H. Hoagland, *Sci, Monthly* **56**, 56 (1943).
7. C. W. Kline, unpublished address before the Society for Scientific Study of Sex, New York (1960).
8. H. J. Muller, in *The Great Issues of Conscience in Modern Medicine* (Dartmouth Medical School, Hanover, N.H., 1960), p. 16.
9. _____ *Sci. Monthly* **37**, 40 (1933).
10. A. Weinstein, *Am. Naturalist* **67**, 222 (1933).
11. This article was originally prepared as a lecture to be given on the invitational program of the Society for the Study of Evolution at its annual meeting in New York, 29 Dec. 1960, but adventitious circumstances prevented its delivery on that occasion.

11
Induction of an Enzyme in Genetically Deficient Rats after Grafting of Normal Liver

ANIL B. MUKHERJEE
AND JOSEPH KRASNER

Abstract. *Tissue from normal rat livers was grafted onto the livers of rats that were genetically deficient in bilirubin uridine diphosphate glucuronyltransferase activity. Twelve weeks after the grafting operation, the liver of the recipient rats had bilirubin uridine diphosphate glucuronyltransferase activity.*

The induction or transfer of enzyme activity would be an attractive mode of therapy for some of the enzyme deficiency diseases. Rugstad and his co-workers (1) have shown that bilirubin uridine diphosphate (UDP) glucuronyltransferase (E.C. 24.1.17) activity could be transferred into enzyme deficient homozygous Gunn rats. This was accomplished by subcutaneous transplantation of a clonal strain of rat hepatoma cells (2). However, transplantation of a viable neoplastic tissue holds little promise of a direct clinical application. In these studies, we grafted small amounts of normal Wistar rat liver into the liver of its enzyme deficient mutant strain, the homozygous Gunn rat, and measured the effects of the grafts upon bilirubin UDP glucuronyltransferase activity and serum bilirubin concentrations in the Gunn rats.

Science, 1973, Vol. 182, Issue 4107, pp. 68-70. Copyright © 1973 by the American Association for the Advancement of Science.

Homozygous Gunn rats were identified by their yellow color shortly after birth. Their ears were punched for subsequent identification. They were weaned at 21 days of age, a blood sample was taken from the orbital sinus cavity, and serum bilirubin concentration was determined by absorbance at 450 nm (Cary model 15 recording spectrophotometer). Elevated bilirubin concentrations in homozygous animals indicated that these rats were deficient in UDP glucuronyltransferase activity. At 6 to 8 weeks, these rats were checked again for serum bilirubin concentration and were used as recipients in the grafting experiments. Donor and recipient rats were matched for age and sex, then simultaneously anesthesized by an intraperitioneal injection of 1 ml of an aqueous solution of tribromoethanol (2 percent). Approximately 5 percent of the liver of each Gunn rat was removed with a uterine punch biopsy forceps. We replaced the punch biopsies of each Gunn rat liver with identical sections from the liver of a normal Wistar rat donor. We transplanted the donor liver from five biopsy sections into each recipient rat liver. Each plug of liver tissue from the Gunn rats was 5 mm in diameter and weighed from 250 to 300 mg. The liver samples from the Gunn rats and the remaining donor livers were put into separate weighed beakers and placed on ice for the subsequent determination of bilirubin UDP glucuro-nyl-transferase activity. Gross morphological and histological changes in the grafted tissue were recorded every 3 weeks for 12 weeks. Four animals were killed sequentially every 3 weeks after surgery. After 12 weeks, the transplanted livers showed no grossly visible trace of the grafted tissue and, histologically, no distortion of structure could be detected.

Rat livers were prepared as 33 percent homogenates as follows. To the weighed livers, two volumes of $0.15M$ KCl solution was added; the mixture was homogenized (Potter homogenizer with Teflon pestle). The homogenates were centrifuged at $9000g$ for 20 minutes at $4°C$ (Sorvall centrifuge); the supernatant was then centrifuged at $104,000g$ for 1 hour (Spinco model L2-65B ultracentrifuge). The pellet which contained the microsomes was suspended in 0.1 M KCl, and homogenized to produce a uniform mixture; bilirubin glucuronyltransferase activity was then assayed (3).

Enzyme activity was measured by addition 0.1 ml of microsomal suspension to 0.15 ml of a solution containing 5×10^{-8} mole of bilirubin (the substrate); 1.5×10^{-5} mole of tris-CHl buffer, pnH 8.0; 2×10^{-7} mole of β-mercaptoethanol; 8×10^{-6} mole of $MgCl_2$ and 5 mg of bovine serum albumin. Both test and control tubes were incubated at $37°C$ for 10 minutes. The reaction was started by adding 0.02 ml of $0.02M$ uridine

TABLE 1. Measurements of bilirubin and UDP glucuronyltransferase activity before and 12 weeks after the graft procedure. We measured UDP glucuronyltransferase activity separately in the left and right lobes of rats 6 and 7. The right lobe received grafts; the left lobe did not receive them.

Rat	Plasma Bilirubin (mg/100 ml)		Enzyme Activity in 30 Minutes (nmole/mg protein)	
	Before	After	Before	After
1	3.8	0.5	0.00	0.17
2	5.9	1.0	0.00	0.54
3	4.8	0.9	0.00	0.38
4	3.7	1.3	0.00	0.66
5	4.2	0.8	0.00	0.55
			Left	*Right*
6	6.0	0.7	0.00	3.77	4.81
7	6.8	0.8	0.00	0.85	1.19

diphosphoglucuronic acid (4×10^{-7} mole) to the test sample and 0.02 ml of water to the control tube. Incubation was continued for 30 minutes in a covered incubator. The entire proceudre was carried out in a darkened room, and tubes containing bilirubin were protected from unnecessary light in order to limit photodegradation of bilirubin. The raction was terminated by the addition of a mixture of ethyl acetate and lactic acid (5:8). The amount of bilirubin glucuronide formed was determined by the method of Weber and Schalm (4).

Total bilirubin in serum from the Gunn rat was determined by measuring absorbance at 450 nm and compared to a standard solution (Dade Reagents obtained from Scientific Products).

Bilirubin UDP glucuronyltransferase activity was lacking in liver microsomes from the homozygous Gunn rat. Serum bilirubin concentrations (4 to 6 mg per 100 ml of serum range) in these animals were relatively high compared to normal rats (less than 1 mg percent). These animals were killed 12 weeks after the grafting operation. Serum bilirubin concentrations were lower, and the liver microsomes showed bilirubin glucuronyltransferase activity (Table 1). Although normal Wistar rats (donors) had a mean enzyme activity of 2.9 nmole of bilirubin glucuronide formed in 30 minutes per milligram of microsomal protein, the individual variation was large. In order to determine whether or not the increased enzyme activity was present in only the grafted liver lobes or throughout the entire liver, we also determined the enzyme activity in the left lobe, which received the normal grafted tissue, and in the right lobe, which did not receive any transplant. No significant differences in enzyme activity in the separate liver lobes were observed (Table 1). In Gunn rats which had received only anesthetic and in those which were sham-operated (punched but no tissue taken out), no significant decrease in serum bilirubin concentrations nor increase in enzymic activity was observed. Transfer of liver from one Gunn rat to another Gunn rat did not alter the enzyme activity, nor did transplanting Gunn rat liver to a normal liver eliminate or decrease enzyme activity below normal levels.

These results demonstrate that it is possible to induce or transfer bilirubin UDP glucuronyltransferase activity in liver microsomes by grafting normal rat liver to Gunn rat liver.

Rugstad *et al.* (5) have shown that rat hepatoma cells have the ability to conjugate bilirubin in vitro. This function is retained after the cells are transplanted subcutaneously into a Gunn rat thereby providing the recipient animal with a normal pathway for bilirubin excretion. However, the enzyme activity is not induced in other tissues. Even if it were, the result would be a hepatoma that would limit the potential for survival of the recipient animals.

We did not investigate the viability or fate of the individual normal liver cells after grafting. Migration and eventual proliferation of selected cells from the normal tissue throughout the recipient Gunn rat liver could account for the appearance of enzyme activity after grafting. It is also possible that a genetic transformation takes place in the deficient liver through incorporation of informational macromolecules from the grafted liver cells. Another possibility is the presence of a stable diffusible derepressor substance in the normal liver tissue which enters the Gunn rat liver cells and enables the totally inactive genetic locus for UDP glucuronyltransferase in the Gunn rat liver to function more efficiently.

The Crigler-Najjar syndrome is the human counterpart of the hyper-bilirubinemia in the Gunn rat (6). There have been several approaches to treatment of this disease, such as barbiturate administration (1, 7, 8) and phototherapy (9) in man and transplantation of functional neoplastic tissue in animals (1) but the effects have been at best only partially beneficial. Our experiments demonstrate a novel biological method for the induction of enzyme activity in the Gunn rat. Although this procedure holds promise for direct clinical application, extrapolation of data from the animal experiments to humans must be cautiously evaluated.

References and Notes

1. H. E. Rugstad, S. H. Robinson, C. Yannoni, A. H. Tashjian, Jr., *Science* 170, 553 (1970).

2. U. I. Richardson, A. H. Tashjian, Jr., F. C. Bancroft, V. I. Richardson, M. B. Goldlust, F. A. Rommel, P. Ofner, *In Vitro* **6**, 32 (1970).

3. E. Halac, Jr., and A. Reff, *Biochim. Biophys. Acta* **139**, 328 (1967); J. Krasner and S. J. Yaffe, *Pediat. Res.* **2**, 307 (1968).

4. A. Ph. Weber and L. Schalm, *Clin. Chim Acta* **1**, 805 (1962).

5. H. E. Rugstad *et al., J. Cell Biol.* **47**, 703 (1970).

6. R. Schmid, in *Metabolic Basis of Inherited Diseases,* J. B. Stanbury, J. B. Wyngaarden, D. S. Frederickson, Eds. (Blakiston, New York, 1966), pp. 881-883.

7. S. H. Robinson, *Nature* **222**, 990 (1969).

8. I. M. Arias, L. M. Gartner, M. Cohen, J. BenEzzer, J. Levi, *Amer. J. Med.* **47**, 395 (1969).

9. R. Gorodischer, G. Levy, J. Krasner, S. J. Yaffe, *N. Engl. J. Med.* **282**, 375 (1970); M. Karon, D. Imach, A. Schwartz, *ibid.,* p. 377.

10. We thank Drs. R. G. Davidson, S. J. Yaffe, and C. Catz for suggestions and criticism. Dr. J. Fisher provided the histopathological interpretations of the grafted liver. We thank M. Cox for technical assistance. Supported in part by grants from the HEW Maternal and Child Health Service (project 417), NIH (HD-05187, HD-04287, HD-06321-01), and the Lalor Foundation.

7 March 1973; revised 24 May 1973

12
Correction of a Genetic Defect in a Mammalian Cell

A.G. SCHWARTZ, P.R. COOK AND HENRY HARRIS

Foreign genetic material, in amounts too small to determine the synthesis of foreign surface antigens, can be incorporated into somatic cells by a special application of the technique of cell fusion. The interpolated genetic material is expressed and replicated in the foreign environment.

When a chick erythrocyte nucleus is introduced into the cytoplasm of a tissue culture cell of the same or a different species, it resumes the synthesis of RNA and DNA[1-3]; and when the reactivated erythrocyte nucleus develops a nucleolus it determines the synthesis of chick specific proteins in the hybrid cell[4,5]. Within 3 or 4 days, the heterokaryons formed by cell fusion enter mitosis and the erythrocyte nuclei can no longer be distinguished as discrete entities. When such heterokaryons enter mitosis, the erythrocyte nuclei are fragmented by a process which has been called "chromosome pulverization" or "premature chromosome condensation"[6]. This process is commonly seen in multinucleate cells when the nuclei show marked asynchrony of DNA synthesis; those nuclei in which the cycle of DNA replication is incomplete when mitosis occurs are usually the ones that are pulverized[7].

These observations suggested that chromosome pulverization might provide a method for incorporating small pieces of foreign genetic

Reprinted from *Nature* March 3, Vol. 230, page 5, 1971.

material into somatic cell nuclei, an operation that has been attempted repeatedly in many different ways, but so far without success. We now report that by using a special application of the cell fusion technique, we can incorporate small amounts of foreign genetic material into somatic cells and that the genes incorporated in this way are expressed and replicated in the foreign environment.

The chick erythrocyte nuclei, obtained from 10 to 12 day old chick embryos, were introduced into the cytoplasm of A_9 cells by the use of inactivated Sendai virus[8]. The A_9 cells are a line of mouse fibroblastic cells deficient in the enzyme inosinic acid pyrophosphorylase[9]. They were selected by continued subcultivation in the presence of 8-azaguanine; cells that have lost inosinic acid pyrophosphorylase cannot incorporate azaguanine into RNA and are thus resistant to the lethal effects of the antimetabolite. When the chick erythrocyte nucleus is reactivated in the cytoplasm of the A_9 cell, it determines the synthesis of inosinic acid pyrophosphorylase[5]; electrophoretic analysis has shown that the inosinic acid pyrophosphorylase formed in these conditions is chick, and not mouse, enzyme[10].

Because the membrane of the mouse nucleus in chick erythrocyte/A_9 heterokaryons undergoes dissolution at mitosis, and the chick nucleus undergoes pulverization, we hoped that a fragment of chick genetic material bearing the structural gene for inosinic acid pyrophosphorylase might be incorporated into the mouse nucleus in some cells during post-mitotic reconstitution. If this were to happen, the resulting cell might be able to continue to produce inosinic acid pyrophosphorylase and thus resist selection against cells that lack this enzyme. The heterokaryons were therefore cultivated in HAT medium, which contains aminopterin, hypoxanthine, thymidine and glycine at concentrations of 4×10^{-7} M, 1×10^{-4} M 1.6×10^{-5} M and 3×10^{-6} M respectively. In this medium, cells lacking inosinic pyrophosphorylase, which therefore cannot synthesize RNA from exogenous sources, do not survive[11]. Cells possessing the enzyme do survive in this medium because the block to endogenous nucleic acid synthesis imposed by the aminopterin can be circumvented by the incorporation of exogenous precursors.

When populations of chick erthrocyte/A_9 heterokaryons were culti- vated in HAT medium, the great majority of the cells gradually died off; but some clones resistant to HAT medium appeared within two or three weeks after cell fusion. When such clones were isolated and populations of cells were grown up from them, these cells were found to contain inosinic acid pyrophosphorylase, and electrophoretic examination of the enzyme by the method of Cook[10] revealed that it migrated in the position characteristic for chick enzyme. This shows that these clones of cells have retained the chick gene for inosinic acid pyrophosphorylase and that this gene continues to be replicated on continued cultivation of the cells *in vitro*.

How Much Genetic Material is Transferred?

When mononucleate hybrid cells are generated from binucleate hetero-karyons, fusion of the two parental nuclei at mitosis usually produces a hybrid nucleus which contains both sets of parental chromosomes[12]. In some cases, many of the chromosomes of one or other parental nucleus are eliminated during the early cell divisions, possibly at the first mitosis[13,14]. We have examined the karyotypes of the HAT-resistant hybrid cells derived from the chick erythrocyte/A₉ heterokaryons as soon as enough cells were generated to permit chromosomal analysis, but we could not recognize any chick chromosomes. Because the chick karyotype contains about fifty very small chromosomes or "micro" chromosomes, whereas the A₉ cell contains only one or two "dot" chromosomes that could be confused with the chick "micro" chromosomes, the presence of the latter in the HAT-resistant hybrid cells should have been obvious.

But the hybrid cells were indistinguishable from A₉ cells in containing only one or two "dot" chromosomes. When the numbers of acrocentric and metacentric chromosomes in the hybrid cells and the A₉ cells were compared, there was no disparity. This shows that the great bulk of the chick genetic material in the heterokaryon had been eliminated from the mononucleate hybrid. But because wild-type A₉ cells show substantial variation in chromosome number, the experiment was repeated on clonal cell populations in which the karyotype was much more stable.

Table 1 compares the chromosomal constitution of a clonal population of hybrid cells carrying the chick gene for inosinic acid pyrophosphorylase with that of a pooled population of cells from which this gene has been lost. These cells were derived from the clonal population of hybrid cells by selection in medium containing 8-azaguanine at a concentration of 4 μg/ml. Cells that produce inosinic acid pyrophosphorylase incorporate the azaguanine into nucleic acids and are thus killed; cells lacking the enzyme survive. The pooled population of cells selected by growth in azaguanine showed a greater variability of karyotype than the clonal population from which it was derived; but an analysis of the two series none the less excludes, with a probability of 0.95, the presence in the cells carrying the

TABLE 1. Mean chromosome numbers of mouse cells with and without chick enzyme

Class of Chromosome	Cells with Chick Enzyme	Cells without Chick Enzyme	95% Confidence Interval
Metacentric	21.76	22.00	−0.52 to +0.07
Metacentric marker 1	1.72	1.37	+0.17 to + 0.55
Metacentric marker 2	0.93	0.81	−0.04 to +0.28
"Dot"	1.78	1.88	−0.29 to +0.06
Acrocentric	28.00	28.86*	−1.20 to −0.50

The means for the cells with chick enzyme were derived from fifty-eight metaphase preparations, the means for the cells without chick enzyme from fifty-two metaphase preparations. The difference between the mean values for each class of chromosome lies, with a probability of 0.95, within the confidence intervals indicated. A maximum value of less than +1.0 in the confidence interval excludes with a probability of at least 0.95 the presence of an additional chromosome in the cells with chick enzyme.

* 104 metaphase preparations counted.

chick gene of even a single additional chromosome in any of the three classes, metacentric, acrocentric or "dot". In the case of the acrocentric chromosomes, the variance does permit the possibility that the cells that have lost the chick gene might contain one more chromosome than the hybrids from which they were derived, but not one less. The presence of a chick "macro" chromosome in the hybrid cells could therefore be accommodated only if the loss of this chromosome in the cells lacking the chick enzyme were compensated by the systematic generation of an additional mouse chromosome, a phenomenon for which it is difficult to envisage a mechanism, or if the cells lacking the chick enzyme were derived only from hybrids that initially contained one chromosome more than the mean, which seems equally implausible. A chick "micro" chromosome could be accommodated only if it were attached in some way to the mouse chromosomes and thus escaped detection.

That very little chick genetic material is present in these hybrid cells is also revealed by tests for the presence of chick-specific antigens on their surface. Species-specific surface antigens can be detected in hybrid cells with great specificity and sensitivity by immunological methods[15]; and the synthesis of chick-specific surface antigens in heterokaryons containing A_9 and chick erythrocyte nuclei has been studied in some detail[4]. Weiss and Green[13] have shown that the genes that determine species-specific antigens are widely distributed in the chromosome set. In man/mouse hybrid cells, human surface antigens could be easily detected even when the hybrid cells retained, on average, fewer than four human chromosomes. Although the mononucleate hybrid cells produced from the chick erythrocyte/A_9 heterokaryons continued to produce chick inosinic acid pyrophosphorylase, no chick-specific antigens could be detected on the surface of these cells by the most sensitive refinement of the mixed haemadsorption reaction. The amount of chick genetic material incorporated into these hybrid cells is clearly very small.

Stability of Incorporated Material

If the chick genetic material were fully integrated into the structure of the mouse chromosomes, the chick genes should not readily be lost from the cell population when the selection pressure against cells lacking inosinic acid pyrophosphorylase is removed. Indeed, if the chick inosinic acid pyrophosphorylase gene were integrated into the chromosome in the same way as the normal mouse inosinic acid pyrophosphorylase gene, the frequency with which inosinic acid pyrophosphorylase activity would be lost from the hybrid cells should be similar to the frequency with which inosinic acid pyrophosphorylaseless mutants are generated in mouse cells cultivated in the same conditions.

We therefore measured the rate of production of cells lacking inosinic acid pyrophorphorylase in populations of the hybrid cells and in populations of L cells (from which the A_9 mutants were originally derived). The two cell populations were first cultivated for 4 months in HAT medium, which selects for the retention of the inosinic acid pyrophosphorylase gene. The selective pressure was then removed by transferring the cells to Eagle's minimal essential medium[16]. The cells were grown in this medium for 6 weeks (about 27 generations) and then transferred to medium containing 8-azaguanine at a concentration of 4 μg/ml., which selects against cells that produce inosinic acid pyrophosphorylase and for cells that lack this enzyme. Analysis of the number of clones resistant to 8-azaguanine in the two populations showed that, after about 75 generations in HAT medium and 27 generations in Eagle's medium, cells lacking inosinic acid pyrophosphorylase appeared in the L cell population at a frequency of less than 1 in 10^6; in the hybrid cell population the frequency was 20%. Direct estimation of enzyme activity in cell homogenates, as described by Harris and Cook[5], confirmed that the cells resistant to 8-azaguanine had indeed lost inosinic acid pyrophosphorylase activity. A comparison of the growth rate, in Eagle's medium, of the hybrid cells containing chick inosinic acid pyrophosphorylase with that of the cells from which chick inosinic acid pyrophosphorylase had been lost, showed that the latter grew slightly less rapidly. The high frequency with which cells lacking inosinic acid pyrophosphorylase appeared in the hybrid cell cultures cannot therefore have been due to these cells having a selective advantage under the cultural conditions used. The high incidence of such cells indicates that, in the absence of selective pressure to retain inosinic acid pyrophosphorylase, the chick inosinic acid pyrophosphorylase gene is lost from the mouse cells with a frequency far exceeding that to be expected by spontaneous mutation.

In other words, the genetic material bearing the chick inosinic acid pyrophosphorylase gene is only loosely integrated into the mouse cells. We do not have decisive evidence about the precise cellular localization of the chick genetic material. It could, in principle, be present in the cytoplasm of the cells; but this seems unlikely for two reasons. First, there is a good deal of evidence that indicates that genetic material introduced artificially into the cytoplasm of somatic cells rapidly undergoes dissolution; and second, if the chick genes were located in the cell cytoplasm it would be difficult to see why only 2 in 10^5 of the heterokaryons (1 in 10^4 of the binucleate heterokaryons) generate mononucleate cells containing the chick genes, because these genes are liberated into the cytoplasm of all heterokaryons when the chick erythrocyte nucleus undergoes pulverization at mitosis. It seems much more probable that the frequency with which mononucleate hybrid cells containing chick inosinic acid pyrophosphorylase are generated from the heterokaryons reflects the frequency with which the chick genetic material bearing the structural gene for this enzyme is incorporated into the mouse nucleus during post-mitotic reconstitution.

Possible Applications of the Technique

Wherever the chick genetic material might ultimately prove to be, our experiments demonstrate that an amount of genetic material too small to determine the synthesis of any species-specific surface antigens can be inserted into a foreign somatic cell, and that the genes resident in this foreign environment can be expressed and replicated. We do not know, although we propose to find out, whether this phenomenon is restricted to a chick nucleus as donor, a mouse nucleus as recipient, or both. We think such stringent restrictions unlikely, for the chromosomes of any nucleus in a multinucleate cell may undergo pulverization if the nucleus has not completed its replication cycle when mitosis occurs. It is therefore not unreasonable to hope that this technique may be generally applicable for inserting small amounts of genetic material into somatic cells. If this is so, there are several possible applications, of which two are perhaps worth mentioning.

If fragments of human genetic material, smaller than a single chromosome, can be inserted into other cells, this would permit human genetic linkage analysis of much greater refinement than is at present possible; and because the interpolated genetic material is rapidly lost on culture of the cells in appropriate selective conditions, the analysis should be a good deal less tedious than the methods currently used. The second application of our finding is more problematical, but it might in the long term prove to be of some practical value. The amelioration of the phenotypic consequences of certain genetic defects in man by the provision of somatic cells not bearing these defects is a therapeutic goal that has often been canvassed as one possible consequence of the activities of those who modify the genetic apparatus of somatic cells. Approaches to this problem are, however, hampered by a number of difficulties, the most fundamental of which is the homograft reaction which eventually ensures the rejection of the foreign cells supplied. If it should prove to be generally possible to correct genetic defects in somatic cells by the insertion of amounts of foreign genetic material too small to determine the synthesis of foreign surface antigens, then the barrier posed by the homograft reaction will have been overcome. For it will then be possible to grow cells from the affected individual, correct the genetic defect in them without introducing foreign surface antigens and then supply the patient with his own cells genetically corrected. There will, of course, still be formidable problems in obtaining the appropriate cells and in administering the corrected cells; but these problems could become the subject of investigation if the homograft barrier were overcome.

This work was supported by the Cancer Research Campaign. A. G. S. is a fellow of the Jane Coffin Childs Memorial Fund, and P. R. C. is Stothert research fellow of the Royal Society.

Received October 14, 1970.

1. Harris, H., *Nature,* **206**, 583 (1965).
2. Harris, H., *J. Cell Sci.,* **2**, 23 (1967).
3. Johnson, R. T., and Harris, H., *J. Cell Sci.,* **5**, 625 (1969).
4. Harris, H., Sidebottom, E., Grace, D. M., and Bramwell, M. E., *J. Cell Sci.,* **4** 499 (1969).

5. Harris, H., and Cook, P. R., *J. Cell Sci.,* **5**, 121 (1969).
6. Johnson, R. T., and Rao, P. N., *Nature,* **226**, 717 (1970).
7. Kato, M., and Sandberg, A. A., *J. Nat. Cancer Inst.,* **41**, 1117, 1125 (1968).
8. Harris, H., and Watkins, J. F., *Nature,* **205**, 40 (1965).
9. Littlefield, J. W., *Nature,* **203**, 1142 (1964).
10. Cook, P. R., J. *Cell Sci.,* **7**, 1 (1970).
11. Szybalska, E., and Szybalski, W., *Proc. US Nat. Acad. Sci.,* **48**, 2026 (1962).
12. Harris, H., *Cell Fusion, The Dunham Lectures* (Clarendon Press, Oxford, 1970).
13. Weiss, M. C., and Green, H., *Proc. US Nat. Acad. Sci.,* **58**, 1104 (1967).
14. Migeon, B. R., and Miller, C. S., *Science,* **162**, 1005 (1968).
15. Watkins, J. F., and Grace, D. M., *J. Cell Sci.,* **2**, 193 (1967).
16. Eagle, H., Oyama, V. I., Levy, M., Horton, C. L., and Fleischman, R., *J. Biol. Chem.,* **218**, 607 (1956).

13
Molecular Biology: Gene Insertion into Mammalian Cells

DANIEL RABOVSKY

.

The problem of inserting specific genes into human cells has intrigued molecular geneticists, and the prospect of the successful solution of this problem has concerned everyone. Both the excitement and the concern have grown now that the armchair speculations—and exploratory results—of a few years ago have matured into hard experimental work. The current results of that work indicate that animal viruses, bacterial viruses, and cell fusion techniques are all capable of introducing new functional genes into mammalian cells, although many of the fundamental genetic and regulatory processes in mammalian cells remain unknown.

Much has been learned about the genetic code and the mechanisms of the replication of DNA, the transcription of DNA into RNA, and the translation of RNA into protein, especially in bacterial cells. A clever and sufficiently industrious molecular geneticist can often produce a specific mutation in any of a large number of genes in the bacterium *Escherichia coli,* can delete genes or add new ones from outside the cell, and can then regulate the expression of genetic traits inside the cell. But the extension of these techniques from bacteria and bacterial viruses (bacteriophages) to nucleated (eukaryotic) cells, especially human cells, awaited new tools and more knowledge.

Several biologists have studied the interaction of foreign DNA with

Science, 1971, Vol. 174, Issue 4012, pp. 933-934. Copyright 1971 by the American Association for the Advancement of Science.

nucleated cells. Among these, Pradman Qasba and Vasken Aposhian at the University of Maryland School of Medicine in Baltimore, have recently shown that one type of animal virus can be used to transport DNA from mouse cells into the nuclei of human cells. At the Roswell Park Memorial Institute in Buffalo, W. Munyon and his co-workers have shown that another type of animal virus may have inserted a specific gene into mouse cells without harming the cells. These workers found that the enzyme specified by this gene was made by the cell and that the new gene seemed to be replicated as the cells divided.

Munyon and his group infected mutant L cells (a line of mouse tissue culture cells) that lacked the enzyme thymidine kinase with the animal virus herpes simplex. The virus had been irradiated with ultraviolet light to decrease its ability to kill cells (1). Herpes simplex virus normally induces a thymidine kinase activity during infection before it kills the cells, but in this experiment about 0.1 percent of the infected L cells were transformed by the irradiated virus into stable cells that had thymidine kinase activity and were maintained in culture for 8 months. No measurable proportion ($< 10^{-8}$) of control L cells gained the ability to express thymidine kinase when uninfected cells or cells infected with a herpes simplex mutant that does not induce thymidine kinase activity were examined.

These results are consistent with the idea that the herpes simplex virus introduced a gene for thymidine kinase into the L cells and that this gene was then maintained and replicated by the cells. However, Munyon notes the possibility that a herpes gene product may have simply induced the stable expression of a gene that was already present in the L cells.

Aposhian has proposed that pseudovirions—which consist of normal virus' protein coats that have enclosed foreign pieces of DNA—might be able to deliver foreign genes into another cell, and that these new genes could function and be replicated in the cell. Qasba and Aposhian have now established that pseudovirions of an animal virus, polyoma, containing labeled DNA from mouse embryo cells can deliver this DNA inside the nuclei of human embryo cells in the form of uncoated pseudovirion DNA as soon as 24 hours after infection of the human cells by the polyoma pseudovirions (2). At present no actual expression or replication of any newly introduced genes has been shown in this system, but many workers think it likely that uncoated mammalian DNA in a mammalian cell nucleus is capable of integrating itself into the host chromosome and functioning in some way.

Carl Merril and his co-workers (3) at the National Institutes of Health in

Bethesda have taken a different approach. They have shown that a bacteriophage is capable of introducing a selected functional gene into human cells, Merril had worked with the bacterial virus lambda phage, one of the transducing phages of *E. coli*. Transducing phages have been among the most useful genetic tools available to molecular biologists. Sometimes bacterial genes are included in the DNA of the new phage produced during infection of a bacterium by a transducing phage. If these phages infect another bacterium they can transduce the bacterial genes that they carry into their new host. In this way transducing phages are often used to selectively introduce new functional genes into bacteria.

The genetic structure of lambda has been very well mapped, and a number of remarkable lambda transducing phages (that is, lambda that carry bacterial genes) are available. Merril's group decided to attempt the transduction of cultured human fibroblast cells from the skin of a patient with galactosemia, the disease that results from an inborn error of metabolism in which the enzyme α-D-galactose-1-phosphate uridyl (GPU) transferase is lacking. The cultured fibroblasts were treated with lambda phage (λ pgal) that carried the *E. coli* galactose operon—a set of genes that codes for the enzymes which convert galactose to glucose and controls their synthesis.

Merril and his co-workers hoped that the GPU transferase gene, which is part of the galactose operon, could be supplied to the fibroblasts by λ pgal and that the fibroblasts would use this transduced gene to make GPU transferase enzyme.

These expectations were fulfilled. Infection of the fibroblasts by λ pgal resulted in both the production of lambda-specific RNA and in the appearance of GPU transferase activity in the fibroblasts. The lambda-specific RNA and the new GPU transferase activity were found in significant amounts and persisted at the same amounts per cell for more than 40 days (during which more than eight doublings of the cells took place). In addition, uncoated λ pgal DNA was, in Merril's experiment, at least as effective as the whole virus particle. Control infections by normal lambda and by λ pgal with a mutation that inactivates its transferase resulted in the production of lambda-specific RNA but not in transferase activity.

There is still no evidence to indicate what part of the cell houses the lambda DNA. But the experiments performed by Merril's group have shown that lambda DNA can enter human cells, and once there at least some of its genes can be replicated, transcribed, and translated.

Cell Fusion Techniques

Viruses are not the only means of introducing new genes into mammalian cells, however. Henry Harris' group at Oxford in England has now shown that cell fusion can also be used to insert a functional gene from chicken cells into mouse tissue culture cells (4). If two cells are fused or if another nucleus is introduced into a cell, the dividing time of the resulting multinucleate cell is primarily determined by the nucleus that is closest to division. The other nuclei may be forced to divide before they are ready, and their chromosomes usually undergo a "premature condensation" which results in their fragmentation or "pulverization."

The Harris group fused chick red blood cells, whose nuclei carry the gene for chick inosinic acid pyrophosphorylase, with mouse fibroblast A_9 cells, which are a mutant line of mouse L cells deficient in this enzyme. During cell division, the nuclear membranes in these hybrid cells disappear, and the chick chromosomes undergo pulverization. Although only mouse nuclei appeared in the daughter cells after mitosis, some (2×10^{-5} of the daughter cells had gained the ability to synthesize chick inosinic acid pyrophosphorylase. However, after more than 100 generations in culture, 20 percent of these transformed hybrid cells lost the ability to make this enzyme, as compared to 1 out of 10^6 in a normal L cell population. Hence the chick gene for inosinic acid pyrophosphorylase was replicated and remained functional in the mouse cells, although it was not as genetically stable as the normal mouse gene for this enzyme. This experiment did not provide direct evidence for the location of the chick gene inside the mouse cells, but indirect evidence leads the Harris group to name the mouse nucleus as its probable residence.

A number of laboratories are already extending these gene transfer techniques. Aposhian's group is now studying the genetic properties of polyoma pseudovirions in mice. They are also attempting to produce polyoma pseudovirions containing the gene for human thymidine kinase. Merril has infected whole animals with transducing lambda bacteriophage in order to determine whether any new traits gained by the animals from the genes that are transported by the phage can be inherited from one generation to the next.

The ability to transfer small segments of DNA from one cell to another gives molecular geneticists the opportunity to map the locations of genes within mammalian chromosomes. In turn, understanding the method by which genes are organized into functional units within the chromosome may reveal how the regulation of gene expression takes place. This regulation may play an important role in the differentiation of animal cells.

Current developments certainly do not yet add up to "genetic engineering"; but there now exists very strong evidence, with a number of different techniques, that experimenters can transfer genes between, and insert them into mammalian cells—a development that opens to experiment many questions about gene function in animals and in humans. Now that a few of the basic tools of molecular genetics have been extended to mammalian cells, more than a few biologists share Aposhian's concern that science "will give us gene therapy before society is prepared for it."

References

1. W. Munyon, E. Kraiselbrud, D. Davis, J. Mann, *J. Virol.* **7**, 813 (1971)
2. P. K. Qasba and H. V. Aposhian, *Proc. Nat. Acad. Sci. U.S.* **68**, 2345 (1971)
3. C. R. Merril, M. R. Geier, J. C. Petricciani, *Nature* **233**, 398 (1971)
4. A. G. Schwartz, P. R. Cook, H. Harris, *Nature New Biol.* 230, 5 (1971).

14
Gene Therapy for Human Genetic Disease?

THEODORE FRIEDMANN
AND RICHARD ROBLIN

At least 1500 distinguishable human diseases are already known to be genetically determined (*1*), and new examples are being reported every year. Many human genetic diseases are rare. For example, the incidence of phenylketonuria is about one per 18,000 live births or about 200 to 300 cases per year in the United States (*2*). Others, such as cystic fibrosis of the pancreas, occur about once in every 2500 live births (*3*). When considered together as a group, however, genetic diseases of humans are becoming an increasingly visible and significant medical problem, at least in the developed countries. While the molecular basis for most of these diseases is not yet understood, a recent review (*4*) listed 92 human disorders for which a genetically determined specific enzyme deficiency has been identified.

Concurrent with the recent progress toward biochemical characterization of human genetic diseases have been the dramatic advances in our understanding of the structure and function of the genetic material, DNA, and our ability to manipulate it in the test tube. Within the last 3 years, both the isolation of a piece of DNA containing a specific group of bacterial genes (*5*) and the complete chemical synthesis of the gene for yeast alanine transfer RNA (*6*) have been reported. These advances have led to proposals (*7*) that exogenous "good" DNA be used to replace the defective DNA in those who suffer from genetic defects. In fact, a first at-

Science, 1972, Vol. 175, Issue 4025, pp. 949-955. Copyright 1972 by the American Association for the Advancement of Science.

tempt to treat patients suffering from a human genetic disease with foreign DNA has already been made (8).

Nevertheless, we believe that examination of the current possibilities for DNA-mediated genetic change in humans in the light of some of the requirements for an ethically acceptable medical treatment raises difficult questions. In order to focus the discussion, in this article we concentrate on the prospects for using isolated DNA segments or mammalian viruses as vectors in gene therapy. For this reason we do not discuss other techniques, such as cell hybridization (9), which have been used to introduce new genetic mateial into mammalian cells. We limit our discussion to the possible therapeutic uses of genetic engineering in humans. The potential eugenic uses, for example, the improvement of human intelligence or other traits, are not discussed because they will be very much more difficult to accomplish (10) and raise rather different ethical questions. Whether genetic engineering techniques can be developed for therapeutic purposes in human patients without leading to eugenic uses is an important question, but lies mostly beyond the scope of this article.

Schematic Model of Genetic Disease

Some aspects of a hypothetical human genetic disease in which an enzyme is defective are shown in Fig. 1. The consequences of a gene mutation which renders enzyme E defective could be (i) failure to synthesize required compounds D and F; (ii) accumulation of abnormally high concentrations of compound C and its further metabolites by other biochemical pathways; (iii) failure to regulate properly the activity of enzyme E_1, because of loss of the normal feedback inhibitor, compound F; and (iv) failure of a regulatory step in a linked pathway because of absence of compounds D or F, as in the increased synthesis of ketosteroids in the adrenogenital syndrome (11). In some cases of human genetic disease, accumulation of high concentrations of compound C and its metabolites appears to do the damage. Often a consequence is mental retardation.

The pathway in Fig. 1 is typical of some recessively inherited genetic defects which result in a deficiency of some gene product, usually an enzyme or hormone. In theory, such defects might be corrected by gene therapy, since such techniques might be able to restore the deficient gene product. Other kinds of genetic defects, including those such as the

Marfan syndrome which show dominant inheritance and those such as Mongolism that are caused by chromosome abnormalities, could probably not be ameliorated by the kind of gene therapy we emphasize here.

Current Therapy

Human genetic diseases are usually treated by dietary therapy (*12*) drug therapy, or gene product replacement therapy (*11*). For example, diets low in lactose or phenylalanine are used as treatments for individuals with galactosemia and phenylketonuria, respectively. Such diets have proved exceedingly effective in galactosemia and have produced a marked reduction in the incidence of mental retardation associated with phenylketonuria. In terms of Fig. 1 this therapy corresponds to restricting the intake of compound A, thus minimizing the accumulation of coumpound C whose further normal metabolism is blocked.

Drug therapy has been used to block or reduce the accumulation of undesired and possibly harmful metabolites. One example is the inhibition of the enzyme xanthine oxidase with the drug allopurinol to reduce the accumulation of uric acid associated with gout and the Lesch-Nyhan syndrome (*3*). At present, this method of treatment has been applied to only a few human genetic diseases. Its more general application clearly depends upon availability of drugs which act selectively on specific enzymes. In another form of drug therapy, drugs which combine specifically with the accumulated compound C are used. An example is the use of D-penicillamine to promote excretion of excess cystine in patients with cystinuria (*11*).

In theory, some human genetic diseases might be alleviated by supplying directly the deficient enzyme (E_3 in Fig. 1). Recently, attempts to treat Fabry's disease (*14*), metachromatic leukodystrophy (*15*), and type 2 glycogenosis (*16*), by adminstering the missing enzyme have been reported. Since exogenous enzyme molecules are eventually inactivated or excreted from the body, repeated enzyme injections would be required to manage the diseases in this way. In time, the patients would probably respond by forming antibodies against the administered enzyme. However, insoluble or encapsulated enzyme preparations may in the future provide a means of supplying therapeutic enzymes in a more stable and perhaps less immunogenic form.

There are growing possibilities for the early detection of some genetic diseases by diagnosis in utero. It is now possible to sample the cells of a

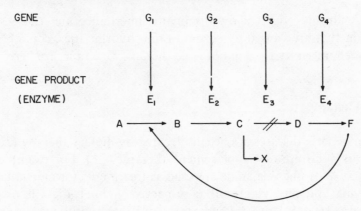

FIGURE 1. A hypothetical pathway for the enzymatic conversion of compound A to a final metabolic product F. Compounds B, C, and D are intermediate products. Four different enzymes, E_1, E_2, E_3, and E_4, the products of the corresponding genes G_1, G_2, and G_4 are required to effect the conversion. The block occurs in the conversion of compound C to D. The concentration of compound F regulates the activity of the first enzyme in the pathway, E_1, in a feedback control loop.

growing fetus in utero and by examining these cells to diagnose a variety of genetic defects (*17*). If genetic defects are detected, some states will permit an abortion if the prospective parents so desire. We recognize that diagnosis in utero and abortion raise difficult social and ethical problems of their own and cite them only to indicate that there are additional alternatives to prospective gene therapy for coping with human genetic disease.

However, many genetic diseases do not yet respond to any of these treatments. For example, most genetic disorders of amino acid metabolism (other than phenylketonuria) cannot be well controlled by dietary therapy. Storage diseases associated with lysosomal enzyme deficiencies (*18*) do not appear to respond to enzyme therapy (*14,15*) and will probably be impossible to control by dietary restriction. In addition, even in cases where disease management is effective, it is seldom perfect. Individuals with diabetes mellitus, when treated with insulin, have an increased incidence of vascular and other disorders and a decreased life expectancy compared to the nondiabetic population (*19*). Children with Lesch-Nyhan syndrome may have their uric acid accumulation controlled by drug (allopurinol) therapy, but their brain dysfunction has, to date, not been reversible.

These limitations of current therapy are stimulating attempts to develop techniques for treating human genetic diseases at the genetic level, the site of the primary defect. Genetic modification of specific characteristics of human cells by means of exogenous DNA seems possible for several reasons. DNA-mediated genetic modification of several different kinds of bacteria has been known for many years (20), and recent experiments suggest that the genetic properties of mutant *Drosophila* strains can be modified by treating their eggs with DNA extracted from other *Drosophila* strains (21). It has also been found that treatment of human cells in vitro with DNA extracted from the oncogenic virus SV40 results in permanent hereditable alteration of several cellular properties (22).

Genetic Modification Mediated by DNA

Permanent, heritable, genetic modification of a human cell by means of DNA required (i) uptake of the exogenous DNA from the extracellular environment; (ii) survival of at least a portion of the DNA during its intracellular passage to the nucleus; (iii) stabilization of the exogenous DNA in the recipient cell; and (iv) expression of the new genes via transcription into an RNA message (mRNA) and translation of this message into the appropriate protein. Some of these processes are illustrated schematically in Fig. 2

Mammalian cells take up proteins, nucleic acids, and viruses from their environment by a process known as endocytosis (23). After binding to the cell membrane, the macromolecules are drawn into the cell membrane leading to vesicle formation (see Fig. 2). Macromolecules contained in vesicles derived from invaginations of the external cell membrane can be degraded if these vesicles fuse with lysosomes. Lysosomes are cell organelles which contain a variety of hydrolytic enzymes. These enzymes can rapidly degrade ingested macromolecules, including DNA (24). Thus, mammalian cells possess mechanisms for protecting themselves from the potentially perturbing influences of foreign DNA.

Despite this cellular defense mechanism, exogenous foreign DNA can, under certain circumstances, become integrated in the DNA of the recipient cell. The evidence of this has come from studies of oncogenic virus transformation of mammalian, including human, cells (25). In the case of oncogenic transformation with SV40 virus, the viral DNA is apparently physically integrated into the chromosomal DNA of the recipient cell (26). It seems probable that heritable alterations of cell morphology and biochemistry are the result of the expression of one or more viral genes. Presumably, viral DNA integration takes place by base pairing of homologous regions of host cell and viral DNA followed by genetic recombination. However, the integration of oncogenic viral DNA may represent a special case since at least one viral gene product may be required for integration (27). Other, nonviral DNA molecules, unable to supply this integration function, might integrate at a much lower frequency, if at all.

In addition to integration by genetic recombination, exogenous DNA might be stabilized in the recipient mammalian cell as an independently replicating genetic unit in the cell nucleus. Although such units are known to exist in bacteria, they have not been observed in mammalian cells. However, the cytoplasmic mitochondria of mammalian cells do contain nonchromosomal, independently replicating units of DNA. The mitochondrial DNA replication system thus offers another possible site for stabilization of exogenous DNA.

For a human genetic defect to be repaired by administering exogenous DNA, the stabilized newly introduced DNA must be correctly expressed. That is, the new gene must be correctly transcribed into mRNA and this mRNA must be correctly translated into protein. Since little is known about the regulation of mRNA synthesis and translation during natural gene expression in mammalian cells, a corresponding high degree of uncertainty exists concerning the ability of newly introduced DNA to be expressed correctly.

A variety of attempts have been made to demonstrate DNA-mediated modification of genetically mutant mammalian cells, both in vivo and in vitro. Apparently successful results in vitro have been reported for diploid human cells lacking the purine "salvage". enzyme hypoxanthine-guanine phosphoribosyltransferase (HGPRT); in human reticulocytes synthesizing an abnormal hemoglobin; in several malignant cells of mouse origin carrying markers for drug resistance; and in mouse cells with defective melanin synthesis, among others (28). In addition, transient expression of HGPRT enzyme function has been detected in human cells deficient in HGPRT after exposing them to DNA from cells with normal amounts of HGPRT (29). This suggests that exogenous DNA may be taken up and expressed, without necessarily being stabilized. However, none of the successful experiments described to date have been reproducible.

There may be several reasons for failure to demonstrate consistently the genetic modification of mammalian cells by DNA. Many previous experiments suffered from the unavailablity of good genetic markers and sensitive selective systems for detecting modified cells. An important difficulty in using bulk DNA isolated from human (or other mammalian) cells is that the fraction of this DNA which is specific for any given gene is estimated to be extremely small, of the order of 10^{-7} (30). As we mentioned earlier, nonviral exogenous DNA may not be able to integrate into the chromosomal DNA of the recipient cell, thus preventing permanent genetic modification.

In spite of the lack of reproducible success in past experiments, several recent technical developments have suggested new ways in which the problems of low DNA specificity, failure of integration, and intracellular DNA degradation might be overcome.

The prospects for directing genetic modification of mammalian cells would almost certainly be enhanced by using DNA preparations containing only the gene for which the genetically defective cells are mutant. As already pointed out, both the isolation of a specific group of bacterial genes and the complete chemical synthesis of a single gene were reported recently ($5,6$).

The RNA-dependent DNA polymerase recently found in RNA tumor viruses (31) could also be used for gene synthesis in vitro. Since this enzyme is able to make DNA copies from an RNA template, it offers a method for synthesizing the DNA for any specific RNA which might be isolated in pure form. Thus, it seems probable that our developing ability to isolate specific genes, or synthesize them, will eventually eliminate the problem of low specificity of the exogenous DNA.

Some workers are developing techniques which could be used to overcome the problem of stabilizing the incoming exogenous DNA in the recipient cell (32). They plan to make use of the ability of the DNA from SV40 virus to integrate into the chromosomal DNA of the cell. Specific genes will be attached to the viral DNA by means of several biochemical steps which are already known and fairly well characterized. These operations would create a hybrid DNA molecule which would carry the information for integration from the original viral DNA and perform the specific gene functions of the attached DNA. In this approach, DNA integration would be combined with biochemical manipulation of the DNA gene substance in vitro, and any gene-specific DNA segments obtained by synthesis or isolation could be utilized. It is clear that before such hybrid DNA molecules could be used in a human therapeutic situation, the oncogenic potential of the viral DNA would have to be eliminated.

In another experimental approach, virus-like particles which contain

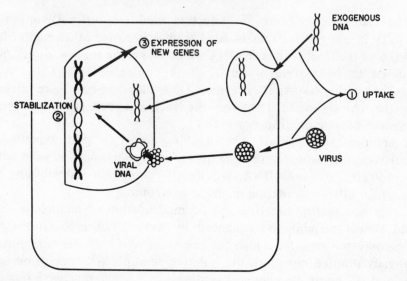

FIGURE 2. Steps in the genetic modification of a mammalian cell. The added exogenous genetic information may be integrated into the chromosome of the recipient cell and become expressed as a new gene product.

pieces of cellular DNA (pseudovirions) instead of viral DNA are being used as the vector for DNA-mediated genetic modification (*33*). This might help to protect the incoming DNA from intracellular degradation. However, pseudovirion DNA is probably a random collection of cellular DNA fragments (*34*) and hence nonspecific for any given gene; it might also be unable to stabilize itself by integration into the host cell DNA. This may explain why attempts to modify thymidine kinase-deficient mouse cells in vitro by means of polyoma virus pseudovirions have been unsuccessful(*35*).

It has recently proved possible to reconstruct infectious particles of several plant and bacterial viruses from the nucleic acid and capsid protein components (*36*). This suggests the possibility of creating artificial pseudoviruses as vectors for DNA-mediated genetic modification. These pseudovirions would contain specific DNA segments (either isolated or synthesized) surrounded by virus capsid protein. The probability of introducing a specific piece of genetic material might be greatly increased when compared with natural pseudovirions carrying randomly excised pieces of DNA. This in no way solves the difficulties of integration and expression of the genetic material. Since the specificity of virus-cell interactions is determined at least in part by the virus capsid protein, encapsidation of specific DNA molecules might confer some cell or tissue selectivity upon the DNA molecules used for gene modification.

DNA-Mediated Gene Therapy

In attempting to envision how DNA might be used as a mediator for the modification of genes in a human being suffering from a genetic defect, we foresee several kinds of new problems. First, the existence of differentiation and cell specialization in the human body will pose several questions. Many human genes are active or expressed only in a small fraction of the cells of the body. For example, the activity of the enzyme phenylalanine hydroxylase (deficient in individuals with phenylketonuria) is demonstrable only in the liver. For prospective gene therapy there might be several consequences. (i) The introduction of, for example, the gene for phenylalanine hydroxylase into cells which do not normally express this enzyme would yield no therapeutic benefits if the expression of the newly introduced genes were also blocked. Methods would have to be developed to deliver the exogenous DNA to the appropriate "target tissue," and to confine its action solely to that tissue. (ii) Some gene products (hormones, for example) are made and secreted by one specialized group of cells and act on target cells elsewhere in the body. Synthesis and secretion of hormones such as insulin are regulated by mechanisms which are still imperfectly understood. Thus, the introduction of new genes for insulin into cells not appropriately differentiated to provide the correct synthetic and secretory responses would be of little use as a treatment for diabetes. (iii) In several genetic disorders, genetic modification of the brain cells themselves may be required to reduce the accumulation of metabolites in the brain, because the blood-brain barrier might prevent enzymes made in other parts of the body from entering the brain (15). We wonder whether direct genetic modification of brain cells could be made safe enough for use in human patients.

Second, regulation of the quantitative aspects of enzyme production may present a problem. By mechanisms as yet unknown, concentrations of cellular enzymes are regulated so that neither too much nor too little enzyme is produced by normal cells. How will we ensure that the correct amount of enzyme will be made from the newly introduced genes? Will the integration event, linking exogenous DNA to the DNA of the recipient cell, itself disturb other cellular regulatory circuits?

Third, the patient's immunological system must not recognize as foreign the enzyme produced under the direction of the newly introduced genes.

If this occurred, the patient would form antibodies against the enzyme protein, perhaps nullifying the intended effects of the genetic intervention. This suggests that the new gene introduced during gene therapy would have to code for an enzyme with the same amino acid sequence as the human enzyme.

In addition, administration of foreign genetic material to patients carries a risk of altering the germ cells as well as the desired target cells. One might think that this problem could be circumvented by first removing some of the patients' cells, carrying out DNA-mediated genetic modification in vitro, and then reimplanting the altered cells back into the patient. However, this approach is likely to be limited by the tendency of cells to dedifferentiate and become malignant when grown in vitro.

For an acceptable genetic treatment of a human genetic defect, we would require that the gene therapy replace the functions of the defective gene segment without causing deleterious side effects either in the treated individual or in his future offspring. Years of work with tissue cultures and in experimental animals with genetic defects will be required to evaluate the potential side effects of gene therapy techniques. In our view, solutions to all these problems are needed before any attempt to use gene therapy in human patients could be considered ethically acceptable.

We are aware, however, that physicians have not always waited for a complete evaluation of new and potentially dangerous therapeutic procedures before using them on human beings. Consider how little was known of the basic aspects of virology during Jenner's development of vaccination against smallpox. In this regard, potential gene therapy techniques resemble other medical innovations. There is currently, and there may continue to be, a tendency to use incompletely understood genetic manipulative techniques, borrowed from molecular biology, in clincial settings. We believe that the first attempt at gene therapy in human patients (8) illustrates this contention.

The case in question (8) concerns two children suffering from hyperargininemia, a hereditary deficiency of the enzyme arginase. The arginase deficiency leads to high concentrations of arginine in the children's blood and cerebrospinal fluid, and has associated with it severe mental retardation (37). An attempt has been made to correct this defect at the genetic level by injecting Shope papilloma virus into the children (8). The scientific rationale for this treatment is based upon the report that the synthesis of arginase is stimulated in rabbit skin infected with Shope papilloma, and that this new arginase activity had some properties which are different from those of the normal enzyme of rabbit liver (38).

In 1958, when these experiments were first reported, it was postulated that the viral DNA carried the gene for a viral arginase different from the cellular enzyme. In addition, the serums of laboratory workers who had worked with and thus been exposed to Shope papilloma virus were tested, and 35 percent of them exhibited lower concentrations of arginine than control hospital patients who had not knowingly been exposed to the virus (39). Thus, there were some grounds for believing that inadvertent infection with Shope papilloma in humans could lower the concentration of serum arginine without apparent harmful effects.

More recently, the interpretation that Shope papilloma virus codes for an arginase has been seriously questioned (40). It now appears more probable that the virus infection stimulates the production of a cellular arginase. Whether the induced arginase is coded for by viral or by cellular genes is important to the rationale of this attempt at gene therapy. If virus infection induces the synthesis of cellular arginase, and if the children have hereditarily lost the ability to produce arginase, then infecting the children with Shope papilloma virus may not have any possibility of correcting their condition (41).

The use of intact viruses as vectors in gene therapy raises further questions. When applied to the skin of rabbits Shope papilloma virus induces skin papillomas, a variable proportion of which develop into cancerous skin lesions. Although Shope papilloma has not had any known harmful effects on humans, tests to establish the safety of large doses have not been performed. It should also be shown that a vector for clinical gene therapy is free from other contaminating viruses latent in the cells used to produce the injected virus.

The clinical results of this therapeutic attempt are not yet known. But we are concerned that this first attempt at gene therapy, which we believe to have been premature, will serve as an impetus for other attempts in the near future. For this reason, we offer the following considerations as a starting point for what we hope will become a widespread discussion of appropriate criteria for the use of genetic manipulative techniques in humans.

Some Preliminary Criteria

We propose the following ethico-scientific criteria which any prospective techniques for gene therapy in human patients should satisfy:

1) There should be adequate biochemical characterization of the prospective patient's genetic disorder. It should be determined whether

the patient (i) is producing a mutated, inactive form of the normal protein; (ii) is producing none of the normal protein; or (iii) is producing the normal protein in normal amounts, but the protein is rendered inactive in some way. For example, alterations in membrane structure leading to loss of the cellular receptors for insulin could produce a diabetes-like condition, even though the patient were producing normal amounts of insulin. We anticipate that defects of this type may be found affecting the activity of enzymes which are normally constituents of cell membranes. Our point is that only in the first type of genetic defect (i) would currently envisioned gene therapy techniques be likely to improve the patient's condition.

2) There should be prior experience with untreated cases of what appears to be the same genetic defect so that the natural history of the disease and the efficacy of alternative therapies can be assessed. Thus, the first reported cases of a new human genetic disease would seldom be candidates for attempts at gene therapy. The reason for this criterion comes from our accumulating experience with some of the better studied genetic defects such as phenylketonuria and galactosemia. We now observe heterogeneity in these conditions; that is, what appears to be the same genetic disease turns out to have different genetic bases in different individuals. Widespread screening for phenylketonuria in newborns has detected individuals who, like phenylketonurics, have high concentrations of phenylalanine in the serum just after birth, but have concentrations in the normal range several months later (2). It is now also clear that some individuals with high concentrations of phenylalanine in the serum have normal intelligence quotients (2). We anticipate that other genetic diseases will exhibit the same kind of heterogeneity. Concern for the welfare of each individual patient dictates that we not rush in with gene therapy until we are very sure about the precise nature and consequences of his genetic defect.

3) There must be an adequate characterization of the quality of the exogenous DNA vector. This will require the development of new, more accurate methods of analyzing the base sequence of the DNA, if synthetic DNA molecules are to be used, or the development of new methods of isolation and purification, if naturally occurring DNA molecules are to be used. We visualize the Food and Drug Administration, or some similar organization, establishing and enforcing quality standards for DNA preparations used in gene therapy.

4) There should be extensive studies in experimental animals to evaluate the therapeutic benefits and adverse side effects of the prospective

techniques. These tests should include long-term studies on the possible induction of cancer and genetic disturbances in the offspring of the treated animals. This will require the development of animal models for human genetic diseases. Previous work, which led to the isolation of a mouse strain deficient in the enzyme catalase (*42*) suggests that such animal models could be developed and might yield answers to some of the questions we have raised.

5) For some genetic diseases, the patient's skin fibroblasts grown in vitro reflect the disorder. Thus, in some cases it would be possible to determine whether the prospective gene therapy technique could restore enzyme function in the cells of the prospective patient. This could be done first in vitro, without any of the risks of treating the whole patient. Some side effects, such as chromosome damage and morphological changes suggesting malignancy, could also be assessed at this time. Only when a potential gene therapy technique had satisfied all these safety and efficacy criteria would it be considered for use in human patients.

These criteria omit some other considerations which we believe are important. Although the ethical problems posed by gene therapy are similar in principle to those posed by other experimental medical treatments, we feel that the irreversible and heritable nature of gene therapy means that the tolerable margin of risk and uncertainty in such therapy is reduced. Physicians usually arrive at a judgment regarding the ethical acceptability of an experimental therapy by balancing the risks and consequences of different available treatments against their potential benefits to the patient. In general, the degree of risk tolerated in medical treatment is directly related to the seriousness of the condition.

High-risk treatments are sometimes considered more justified in life-threatening situations. For different human genetic diseases, the severity of the problem in the untreated condition and the response to currently available therapy varies greatly. Thus, phenylketonuria leads to mental retardation, but not death, in most untreated affected individuals, but the mental retardation can be avoided for the most part by prompt neonatal dietary therapy. In contrast, in the infantile form of Gaucher's disease, a deficiency in the enzyme glucocerebrosidase (important in the metabolism of brain glycolipids) leads to severe and progressive neurologic damage and death within 1 or 2 years (*38*). There is as yet no effective therapy. Thus, the specific characteristics of each genetic disease will be an important factor in evaluating whether or not to attempt gene therapy. We believe that the prospective use of gene therapy will need to be evaluated on a case by case basis.

Another ethical ideal which guides experimental medical treatments is informed consent. By informed consent we mean that the patient, after having the nature of the proposed treatment and its known and suspected risks explained to him by the physician, freely gives the physician his consent to proceed with the treatment. Since many of the cases where gene therapy might be indicated will involve children or newborns as patients, there will be especially troubling problems surrounding informed consent. Parents of newborn children with genetic defects may be asked to give "consent by proxy" for gene therapy. Clearly, until we know much more about the side effects of gene therapy, it will not be possible to provide them with adequate information about risks to the treated individual and his offspring.

Control of Gene Therapy

How can gene therapy in humans be controlled to avoid its misuse? By misuse we mean the premature application of techniques which are inadequately understood and the application of gene therapy for anything other than for the primary benefit of the patient with the genetic disease. In our view, it will be possible to control the procedures used for gene therapy at several levels. For example, between the patient and physician, we can usually rely upon the selection of a therapeutic technique having optimal chances of success. In general, we believe that the doctor will not recommend and the patient will not accept an uncertain, risk-laden gene therapy if a reasonably effective alternative therapy is available. However, the physician, in this as in other cases of experimental therapeutic techniques, has a near monoply on the relevant facts about risks and benefits of various treatments. Since the physician concerned may also be active in trying to develop the gene therapy technique, how can the patient be protected from a physician who might be over-eager to try out his new procedure?

It seems to us that significant opportunities for control also exist at the level of the hospital committees responsible for examining experimental techniques. Already at accredited hospitals, all proposals for research in which human subjects will be used must pass through a review committee. Further control exists through scrutiny of the proposed techniques by the physician's immediate peers.

Procedures to be used for gene therapy might also be controlled by the committees and organizations approving and funding research grants.

Moderately large amounts of money will be required for the development of gene therapy techniques, hence there should be competition for public funds with other urgent medical needs. Thus, the first use of gene therapy in human patients would, of necessity, have secured the implied or direct approval of several larger public bodies beyond the principal physician-investigator. In our judgment, these levels of control will probably prove adequate to prevent misuse of projected gene therapy if, as we suspect, gene therapy is attempted in only a small number of instances. Any potential large-scale use of gene therapy (for example, the prospect of treating the approximately 4 million diabetics in the United States with DNA containing the gene for insulin) might appreciably affect the overall quality of the gene pool and would require other forms of control.

Conclusions

In our view, gene therapy may ameliorate some human genetic diseases in the future. For this reason, we believe that research directed at the development of techniques for gene therapy should continue. For the foreseeable future, however, we oppose any further attempts at gene therapy in human patients because (i) our understanding of such basic processes as gene regulation and genetic recombination in human cells is inadequate; (ii) our understanding of the details of the relation between the molecular defect and the disease state is rudimentary for essentially all genetic diseases; and (iii) we have no information on the short-range and long-term side effects of gene therapy. We therefore propose that a sustained effort be made to formulate a complete set of ethico-scientific criteria to guide the development and clinical application of gene therapy techniques. Such an endeavor could go a long way toward ensuring that gene therapy is used in humans only in those instances where it will prove beneficial, and toward preventing its misuse through premature application.

Two recent papers have provided new demonstrations of directed genetic modification of mammalian cells. Munyon et al. (44) restored the ability to synthesize the enzyme thymidine kinase to thymidine kinase-deficient mouse cells by infection with ultraviolet-irradiated herpes simplex virus. In their experiments the DNA from herpes simplex virus, which contains a gene coding for thymidine kinase, may have formed a hereditable association with the mouse cells. Merril et al. (45) reported that treatment of fibroblasts from patients with galactosemia with exogenous DNA caused increased activity of a missing enzyme, α-D-galactose-1-phosphate uridyltransferase. They also provided some evidence that the change persisted after subculturing the treated cells. If this latter report can be confirmed, the feasibility of directed genetic modification of human cells would be clearly demonstrated, considerably enhancing the technical prospects for gene therapy.

References and Notes

1. V. A. McKusick, *Mendelian Inheritance in Man* (Johns Hopkins Press, Baltimore, ed. 3, 1971).
2. D. Y. Y. Hsia, *Progr. Med. Genet.* **7**, 29 (1970).
3. E. R. Kramm, M. M. Crane, M. G. Sirken, M. D. Brown, *Amer. J. Public Health* **52**, 2041 (1962).
4. V. A. McKusick, *Annu. Rev. Genet.* **4**, 1 (1970).
5. J. Shapiro, L. MacHattie, L. Eron, G. Ihler, K. Ippen, J. Beckwith, *Nature* **224**, 768 (1969).
6. K. L. Agarwal, H. Buchi, M. H. Caruthers, N. Gupta, H. G. Khorana, K. Kleppe, A. Kumar, E. Ohtsuka, U. L. Rajbhandary, J. H. Van De Sande, V. Sgaramella, H. Weber, T. Yamada, *Ibid.* **227**, 27 (1970).
7. S. Rogers, *New Sci.* (29 Jan. 1970), p. 194; H. V. Aposhian, *Perspect. Biol. Med.* **14**, 98 (1970).
8. *New York Times*, 20 Sept. 1970.
9. A. G. Schwartz, P. R. Cook, H. Harris, *Nature* **230**, 5 (1971).
10. B. D. Davis, *Science* **170**, 1279 (1970).
11. J. B. Stanbury, J. B. Wyngaarden D. S. Fredrickson, Eds., *The Metabolic Basis of Inherited Disease* (McGraw-Hill, New York, ed. 2, 1966).
12. N. A. Holtzman, *Annu. Rev. Med.* **21**, 335 (1970).
13. W. N. Kelley, F. M. Rosenbloom, J. Miller, J. E. Seegmiller, *N. Engl. J. Med.* **278**, 287 (1968).
14. C. A. Mapes, R. L. Anderson, C. C. Sweeley, R. J. Desnick, W. Krivit, *Science* **169**, 987 (1970).
15. H. L. Greene, G. Hug, W. K. Schubert, *Arch. Neurol.* **20**, 147 (1969).
16. G. Hug and W. K. Schubert, *J. Cell Biol.* **35**, C1 (1967).
17. A. Milunsky, J. W. Littlefield, J. N. Kanfer, E. H. Kolodney, V. E. Shih, L. Atkins, *N. Engl. J. Med.* **283**, 1370, 1441, 1498 (1970).
18. R. O. Brady, *Annu. Rev. Med.* **21**, 317 (1970).
19. S. Pell and C. A. D'Alonzo, *J. Amer. Med. Ass.* **214**, 1833 (1970).
20. R. D. Hotchkiss and M. Gabor, *Annu. Rev. Genet.* **4**, 193 (1970).
21. A. S. Fox, S. B. Yoon, W. M. Gelbart, *Proc. Nat. Acad. Sci. U.S.* **68**, 342 (1971).
22. S. A. Aaronson and G. J. Todaro, *Science* **166**, 390 (1969).
23. H. J. P. Ryser, *ibid.* **159**, 390 (1968).
24. J. T. Dingle and H. B. Fell, *Frontiers Biol.* **14A**, 220 (1969).
25. R. Dulbecco, *Science* **166**, 962 (1969); M. Green, *Annu. Rev. Biochem.* **39**, 701 (1970).

26. J. Sambrook, H. Westphal, P. R. Srinivasan, R. Dulbecco, *Proc. Nat. Acad. Sci. U.S.* **60**, 1288 (1968).

27. M. Fried, *Virology* 40, 605 (1970).

28. E. H. Szybalska and W. Szybalski, *Proc. Nat. Acad. Sci. U.S.* **48**, 2026 (1962); M. Fox, B. W. Fox, S. R. Ayad, *Nature* 222, 1086 (1969); R. A. Roosa and E. Bailey, *J. Cell. Physiol.* **75**, 137 (1970); L. M. Kraus, *Nature* **192**, 1055 (1961); D. Roth, M. Manjon, M. London, *Exp. Cell Res.* **53**, 101 (1968); J. L. Glick and C. Sahler, *Cancer Res.* **27**, 2342 (1967); E. Ottolenghi-Nightingale, *Proc. Nat. Acad. Sci. U.S.* **64**, 184 (1969).

29. T. Friedmann, J. H. Subak-Sharpe, W. Fujimoto, J. E. Seegmiller, paper presented at Society of Human Genetics Meeting, San Francisco Oct. 1969.

30. Calculated with the assumption that (i) the DNA contents of a diploid human cell is 4×10^{12} daltons; (ii) representative gene codes for a protein containing 200 amino acids equivalent to 4×10^5 daltons of DNA; and (iii) there are two copies of each gene per cell. This would represent a minimum estimate if the redundant DNA segments in human cells include genes which specify enzymes.

31. H. M. Temin and S. Mizutani, *Nature* **226**, 1211 (1970); D. Baltimore, *ibid.*, p. 1209.

32. P. Berg, and D. M. Jackson, personal communication.

33. J. V. Osterman, A. Waddell, H. V. Aposhian, *Proc. Nat. Acad. Sci. U.S.* **67**, 37 (1970).

34. L. Grady, D. Axelrod, D. Trilling, *ibid.*, p. 1886.

35. B. Hirt, seminar at Brandeis University, Oct. 1970.

36. H. Fraenkel-Conrat, *Annu. Rev. Microbiol.* 24, 463 (1970).

37. H. G. Terheggen, A. Schwenk, A. Lowenthal, M. Van Sande, J. P. Colombo, *Lancet* **II-1969**, 748 (1969); *Z. Kinderheilk* 107, 298 and 313 (1970).

38. S. Rogers, *Nature* 183, 1815 (1959).

39. _____ and M. Moore, *J. Exp. Med.* 117, 521 (1963).

40. P. S. Satoh, T. O. Yoshida, Y. Ito, *Virology* 33, 354 (1967); G. Orth, F. Vielle, J. B. Changeux, *ibid.* 31, 729 (1967).

41. This objection would be straightforward if only one type of arginase were present in human cells. Even though there seem to be two arginase isozymes in both human liver and erythrocytes, the objection is still cogent since one isozyme contributes 90 to 95 percent of the total arginase activity, and the two isozymes cross-react immunologically. See J. Cabello, V. Prajoux, M. Plaza, *Biochim. Biophys. Acta* 105, 583 (1965).

42. R. N. Feinstein, M. E. Seaholm, J. B. Howard, W. L. Russel, *Proc. Nat. Acad. Sci. U.S.* **52**, 661 (1964).

43. H. Harris, *Frontiers Biol.* 19, 167 (1970).

44. W. Munyon, E. Kraiselburd, D. Davis, J. Mann, *J. Virol.* 7, 813 (1971).

45. C. R. Merril, M. R. Geier, J. C. Petricciani, *Nature* 233, 398 (1971).

15
Embryology: Out of the Womb – into the Test Tube

JEAN L. MARX

The possibility of "creating" human life in the test tube has long fascinated—or repelled—scientists and laymen alike. Because of the progress in manipulating embryos in vitro, many investigators now think that this possibility, with its attendant ethical, moral, and legal dilemmas (see box), will be realized in the near future. Techniques for in vitro fertilization of human eggs, as well as the eggs of common laboratory animals such as mice, rabbits, and hamsters, are available. Moreover researchers are able to grow mammalian embryos in culture and even to freeze them.

These advances are significant, not just because of their potential application to human reproduction, but also because they permit study of fundamental problems of genetics and development. Furthermore, if applied to the highly practical realm of animal husbandry, they could facilitate the breeding of superior cattle.

The success of the molecular biologists in explaining control of genetic expression in bacterial cells has yet to be translated into a similar understanding of these processes in the cells of developing mammalin embryos. The study of biochemical and physiological events in embryos developing within the uterus of a living animal presents formidable difficulties. The alternative—studying embryo development in vitro—has

Science 1973, Vol. 182, Issue 4114, pp. 811-814. Copyright 1973 by the American Association for the Advancement of Science.

been hindered by lack of culture systems in which embryonic cells would differentiate normally throughout the entire gestational period.

Recently, however, Yu-Chih Hsu of the Johns Hopkins University School of Hygiene and Public Health, Baltimore, Maryland, devised a system for culturing mouse embryos from the blastocyst stage to a state of development approximately equivalent to that seen after 9 days of gestation in vivo—or almost half of the 21-day gestation period of the mouse. Previous investigators were able to culture fertilized mouse eggs only to the blastocyst stage or to culture embryos older than 7.5 days. (The blastocyst is a hollow ball of cells that forms, in the mouse, 3 to 4 days after fertilization; at the end of this stage—5 days after fertilization— the embryo implants in the uterine wall.) Thus, Hsu bridged a gap in embryo culture that includes the critical period during which implantation occurs.

The requirement for blastocysts to implant before undergoing further development may account for earlier failures to culture embryos beyond the blastocyst stage. Hsu overcame this obstacle by using petri dishes coated with reconstituted rat tail collagen to serve as a substrate for implantation. The embryos did attach to the collagen and develop normally.

According to Hsu, the kind of serum added to the culture medium is critically important to embryo development. He uses calf serum for the first and second days of culture, fetal calf serum through the fourth day; a mixture of fetal calf and human cord serums on the fifth and sixth days; and, finally, human cord serum. He is trying to identify the serum components that are active in the different stages of development.

Most of Hsu's investigations have been performed with embryos resulting from in vivo fertilization. Female mice are treated with hormones that stimulate egg maturation and release (a process called "superovulation" that resembles the hormone treatments producing multiple births in humans) and then mated. After 35 days of gestation the embryos are washed out of the oviducts and cultured. Hsu now says that he has cultured eggs fertilized in vitro; they also attained a state of development equivalent to that after about 9 days of gestation in vivo.

A number of investigators have studied the metabolism of mammalian embryos and the conditions needed for their culture. Included among them are John D. Biggers, Harvard University Medical School, Cambridge, Massachuestts; Ralph Brinster, University of Pennsylvania School of Veterinary Medicine, Philadelphia; Wesley K. Whitten, Jackson Laboratory, Bar Harbor, Maine; and David G. Whittingham, University of

Cambridge, Cambridge, England. In general, embryo culture requires carefully controlled conditions of temperature (37°C) of pH (around 7.4), carbon dioxide, high humidity, an ionic composition similar to serum, a source of amino nitrogen (frequently bovine serum albumin), and an energy source.

The energy requirements of embryos change during gestation. Whitten found that one- or two-cell embryos would not divide with glucose as an energy source. Brinster later showed that pyruvate is the principal energy source for supporting development of such early embryos. The eight-cell stage, however, could use glucose and a number of other compounds. By the time of implantation, energy metabolism of embryonic cells resembles that of other cells.

Brinster thinks that there may be considerable similarity between preimplantation embryos of different mammalian species. He points out that before implantation their development follows parallel courses in both morphology and timing of development, although total gestational periods may vary greatly. Consequently, much of the current information, usually derived from studies on mice or rabbits, may also apply to other species, including humans.

Embryos can be maintained in culture for relatively short times, but they cannot be stored for prolonged periods as is frequently desirable. For example, breeding colonies must now be maintained—a process both time-consuming and expensive—in order to preserve rare strains of mutant mice even though they are not presently needed for research. Last year, Whittingham, in collaboration with Peter Mazur and Stanley Leibo at Oak Ridge National Laboratory, Oak Ridge, Tennessee, found that mouse embryos could be frozen at temperatures as low as −269°C for as long as 8 days—and survive. Some investigators think that the ability to store frozen embryos might eliminate or reduce the need for maintaining colonies of animals not in use. Since it is possible to transplant mouse embryos into foster-mothers in which they will develop into newborn mice, the frozen embryos could be thawed when needed and grown to term in foster mothers.

Whittingham, Mazur, and Leibo used two criteria of survival for their frozen embryos—development to the late blastocyst in culture and development to living mice in the uteri of foster-mothers. Up to 70 percent of the embryos frozen in the one-, two-, or eight-cell or blastocyst stage fulfilled the first criterion. Almost 1000 of the thawed embryos were subsequently transplanted into foster-mothers. Sixty-five percent of the animals became pregnant. Forty-three percent of the transplanted embryos developed into living fetuses (killed after 18 days of gestation) or newborn, apparently normal mice. The fetuses or pups carried genetic markers—dark eyes and coats—not possessed by their albino foster-mothers. Mazur says that they have now frozen mouse embryos for periods up to 1 year with survival of 80 percent of the thawed embryos.

According to Mazur and Leibo, who are specialists in cryobiology rather than embryology, formation of ice crystals within cells during freezing usually produces irreversible damage. In their experiments with Whitting-ham, however, they avoided ice crystal formation by employing extremely slow rates of cooling—0.3° to 2°C per minute—to allow enough time for freezable water to flow out of the cells. They also found that slow warming—4° to 25°C per minute—was necessary for survival. A third requirement was the addition of a protective agent that helps prevent freezing damage, possibly through action on cell membranes.

In addition to using frozen embryos to preserve mutant strains of laboratory animals, or even of endangered species, they may also be useful for transportation of animals. Whitten, for example, recently shipped frozen mouse embryos to Whittingham in England. Whitten said that 57 embryos were implanted into foster-mothers; 21 developed into fetuses (killed before parturition) and 11 into newborn mice. Live animals require special handling during shipment and may also be subjected to quarantine against infectious disease if transported across national borders. Frozen embryos are much less likely to carry diseases, such as hoof-and-mouth disease or rabies, than are adults.

Large animals are particularly difficult to transport. Cattle breeders have been importing from Europe "exotic" breeds of beef cattle for improving their herds. Embryos frozen for transport or storage, or even for preservation of unpopular breeds, would facilitate the breeding of superior cattle and the practice of animal husbandry in general. In fact, the Agricultural Research Council's Unit of Reproductive Physiology and Biochemistry, Cambridge, England, was midwife to the birth, in June of this year, of Frosty, a bull that developed from a deep-frozen embryo. The Cambridge group, under the direction of Ian Wilmut and L. E. A. Rowson, transplanted, into 11 foster-mothers, 21 embryos that had been frozen at −196°C and then thawed. Two implanted into the uterus of the same recipient, but only one survived to term. Thus, the process, while apparently feasible, is not yet ready for routine application.

Because of growing demands for beef, animal scientists are examining ways to increase production by increasing the number of high-quality animals. Most cows can produce only one calf per year, or six or seven in a lifetime. This number can be increased by transplanting embryos from superior breeding animals to healthy but otherwise undistinguished foster-mothers. The donor is first "superovulated" to stimulate egg release from her ovaries. This can provide up to 30 eggs (instead of just one) to be fertilized by artificial insemination of the donor. The embryos are removed surgically and transplanted to the foster-mother. The recipient must be in the stage of estrus when implantation can occur.

The process may appear expensive, but is well within the limits of economic feasibility. A number of companies provide the service commercially. According to Casey Ringelberg of Modern Ova Trends, Norval, Ontario, the average fee for each calf produced by this procedure is $2500 to $3000; however, breedable heifers of breeds like Limousin, Simmental, or Chianina sell for $20,000 to as much as $100,000 per animal—and the breeder gets a tax break because he can deduct the cost of producing the animal as a business expense.

Ringelberg points out that the efficiency of the process could still be increased—for example, by freezing embryos for transport or storage until the recipient was in the right stage of estrus. A major handicap is the lack of an in vitro fertilization technique for bovine eggs. Despite successes with common laboratory animals—and even with the human—in vitro fertilization of bovine eggs has not yet been accomplished. Such a technique would mean that eggs and sperm, collected from donors anywhere in the world, could be used to prepare embryos that could then be transplanted into foster-mothers.

In vitro fertilization of eggs from other species is now done routinely in several laboratories. Although investigators had tried to fertilize mammalian eggs in vitro for almost a century, progress was slow until sperm capacitation was discovered in the early 1950's. Capacitation is a still poorly understood process that sperm undergo in the female reproductive tract before they can penetrate the egg.

In Vitro Fertilization

Development of in vitro fertilization techniques enabled investigators to study the mechanism of fertilization in systems much simpler than the living animal. For example, M. C. Chang of the Worcester Foundation, for Experimental Biology, Shrewsbury, Massachusetts, has studied fertilization, both in vitro and in vivo, in the mouse, hamster, rabbit, and rat. In addition to defining conditions necessary for fertilization in vitro, Chang has used in vitro conditions to capacitate sperm. He found that the sperm of some species could be capacitated in media with well-defined compositions that did not include fluids from the female reproductive tract. A medium containing bovine serum albumin gave good capacitation of mouse sperm.

Capacitation of rabbit sperm appears to have more stringent requirements. However, Benjamin Brackett of the University of Pennsylvania School of Medicine, Philadelphia, found that he could achieve this by incubating the sperm with uterine fluid before using them for in vitro fertilization. More recently, Brackett, in collaboration with Gene Oliphant of the University of Virginia, Charlottesville, found that increasing the ionic strength of the incubation medium can produce capacitation of rabbit sperm in the absence of uterine fluid.

Brackett says that it is frequently difficult to select suitable criteria to prove that a sperm has actually fertilized an egg. Some eggs can divide, when appropriately stimulated, even though they have not been fertilized (a process called parthenogenesis) or they may undergo degenerative changes that resemble those of a fertilized egg. The ultimate criterion is transplantation of the resulting embryo to a foster-mother and its subsequent development to a fetus. This criterion has been satisfied for the mouse and rabbit, but not for the human—although it has been tried.

Several investigators have reported in vitro fertilization of human eggs. Among these are R. G. Edwards of the University of Cambridge and Patrick Steptoe of Oldham General Hospital, Oldham, England. They have also attempted to transplant the resulting embryos into the uteri of women (the donors of the eggs) and to bring them to term. So far, no such "pregnancy" has lasted longer than 21 days. Edwards and Steptoe believe that the principal problems are in the implantation technique and in duplicating the correct hormonal conditions for implantation. Carl Wood of Monash University, Victoria, Australia, reported doing a similar experiment, also without success.

The goal of these investigators is to circumvent a certain type of infertility. Some women have healthy ovaries and a uterus but cannot conceive because their oviducts are blocked or defective. But if her eggs are removed by laparoscopy, fertilized in vitro with her husband's sperm, and returned to her uterus, such a woman may be able to bear her own child.

Although Brackett has attempted in vitro fertilization of human eggs, he believes that additional experimentation on laboratory animals, especially non-human primates, is desirable. Not only would this allow time to clarify and solve the biological problems but also to confront the bioethical issues. Brackett is studying in vitro fertilization of rhesus monkey eggs but progress has been slow.

A better understanding of the mechanism and requirements of fertilization can be applied to prevention of conception as well as to treatment of infertility. It may be possible to interfere with sperm capacitation, or otherwise alter conditions within the female reproductive tract and prevent union of sperm and egg. Whatever its applications, current research on fertilization and development impinges on the most profound questions of life and birth.

In Vitro Fertilization of Human Eggs: Bioethical and Legal Considerations

Is in vitro fertilization of human eggs a valid means for alleviating the suffering of infertile couples? Or is it a dehumanizing and illicit intrusion of technology into one of the most profound aspects of human life? These are among the bioethical questions raised by current research in human embryology.

One rationale for much of this research is that it may permit a married couple, infertile because the woman's oviducts are defective, to have their own biological child. The woman could bear the child if an embryo, obatained from in vitro fertilization of her eggs by her husband's sperm, could be implanted in her uterus to develop.

The stumbling block for many who object to this procedure is the unknown element of risk to the embryo—destined to become a human being if pregnancy results. Paul Ramsey of the Department of Religion, Princeton University, Princeton, New Jersey, thinks that in vitro fertilization constitutes unethical medical experimentation on potential human beings. He argues that it is possible to exclude the possibility that manipulations performed on the embryo outside the womb will damage it and result in production of a deformed or handicapped human. Experiments on animals might prove that the techniques of in vitro fertilization, embryo culture, and implantation were safe for animals. But only human experimentation could prove them safe for the human—and, according to Ramsey, that experimentation is unethical because of the risk involved.

Ramsey thinks that detection of embryo damage is not the answer to this dilemma. Before implantation, some abnormalities may escape detection or the detection methods themselves may be harmful; after implantation, amniocentesis (sampling of uterine fluids plus fetal cells) could identify some defects so that an abortion could be performed, but amniocentesis itself entails an added element of risk.

At present, no one knows whether in vitro manipulations of the embryo involve greater or less risk than does ordinary conception in vivo. But it is known that all conceptions are fraught with risk. Marc Lappé of the Institute of Society, Ethics, and the Life Sciences, Hastings-on-Hudson, New York, pointed out that up to 20 percent of human pregnancies, usually those in which the fetus is abnormal, may abort spontaneously. He also believes that research on the early mammalian embryo during the preimplantation stages indicates that it is quite resistant to damage. Most known teratogenic effects, for example, appear to occur after implantation during organ development.

Lappé thinks that in vitro fertilization can be a legitimate means of fulfilling a couple's desire for a child, if the risks prove to be acceptable and if the parents understand and consent to them. Nevertheless, he has suggested a moratorium on human research until animal experimentation, especially on subhuman primates, demonstrates that the risks of in vitro fertilization are at least no greater than those of normal conception.

Leon Kass, a scientist who frequently writes on bioethical topics, questions whether, in an age when over-population is a major concern, there are compelling reasons to proceed rapidly with the development of new means of producing babies—especially since, in his view, these techniques would introduce elements of depersonalization and dehumanization into the act of human procreation. Kass points out that there is an alternate solution to this problem of infertility, an operation for reconstruction of the oviducts. He suggests that additional effort be expended to improve the operation, now frequently unsuccessful, because it can cure the defect that causes the infertility without raising complex ethical issues.

Risk to a potential human being is not the only basis for objections to in vitro fertilization. To some individuals, the termination of fetal life—by whom, at what stage of development, by what method—is one of the central bioethical problems. For embryos would undoubtedly be killed, even in cases where implantation is the goal. More than one egg is fertilized, but only one is implanted.

Although there are analogies to abortion, Kass suggests that the issues involved in the two cases differ in part. Embryos produced by in vitro fertilization are wanted, used, and then deliberately killed; embryos that are aborted are usually the result of an "accidental" conception. Furthermore, André Hellegers, Director of the Kennedy Institute for the Study of Human Reproduction and Bioethics, Washington, D.C., points out that the basis for allowing abortion is conflict between the woman and fetus; the fetus invades her privacy or is a threat to her physical or mental health. For embryos in vitro there can be no such conflict.

Use of in vitro fertilization techniques need not be restricted to treating infertility. They could also be applied to eugenics. This could either be positive eugenics, breeding "superior" human beings by mating eggs and sperm from donors with the desired qualities; or negative eugenics, discarding embryos carrying genetic defects (if these can be detected). Most investigators agree that the consequences of such tampering with human evolution—itself poorly understood—are unknown. The same could be said of the possibility of using these techniques to allow parents to predetermine the sex of their children. All of these applications require destruction of some embryos.

Experiments in human genetics and development could be performed on embryos obtained from in vitro fertilization. For example, controlled mating experiments in the human are not possible. The scientific difficulties (such as long generation time and small number of progeny) are almost as insurmountable as the ethical ones. However, such mating could be achieved in the test tube and the resulting embryos studied in the initial stages of development for expression of a genetic trait such as synthesis of an enzyme. One scientist suggested that these embryos not be maintained beyond the blastocyst stage. Some do not believe that these early embryos—which are barely visible to the naked eye and have not yet differentiated—are human life worthy of protection; others do, however.

Underlying any discussion of bioethics and the legality or morality of scientific research is still another thorny problem—one about which members of the scientific community are highly sensitive. That is the question of regulation of research. Who decides what is permissible and how is the decision enforced? In the case of research funded by the federal government, enforcement, at least, is relatively simple: deny funds for research not meeting the required guidelines.

When Robert Marston was director of the National Institutes of Health (NIH), Bethesda, Maryland, he commissioned an NIH task force, unofficially known as the Human Investigations Committee, to study the ethical and legal issues of human experimentation and to recommend guidelines for NIH-supported research. The committee report, titled "Draft: Special policy statement on the protection of human subjects involved in research, development, and demonstration activities" was published in the *Federal Register* on 16 November.

A subcommittee of the task force probed (among other things) questions relating to the use of human fetal material, whether derived from fertilization in vivo or in vitro, in research, A scientist who served on the subcommittee outlined some of their considerations on the vitro fertilization. (He requested that his name not be used to prevent a deluge of what he characterized as hate mail.) The scientist pointed out that there were legal issues in addition to moral or ethical issues.

Unlike natural conception, in vitro fertilization requires participation by a third party—the investigator or physician who fertilizes and implants the egg in the recipient. If a defective child develops from the egg, would the third party be legally liable for damages? And would the agency that funded the research be liable? At present NIH prohibits investigators with NIH funding from requiring participants in their research projects to sign waivers that release the institution from liability for damages. Even if a participant did sign such a waiver, it would not prevent him from suing for damages.

Furthermore, the question of who is legally responsible for caring for the child must be considered. The simplest case is that in which the couple is married; the husband donates the sperm, and the wife donates the egg and carries their child. But the use of in vitro fertilization need not be restricted to the simplest case. The transplant recipient and egg donor may not be the same individual. A man other than the husband may donate the sperm. A number of variations are possible. And finally there is the possibility, however remote, that the embryo can be brought to term completely in vitro. Will the child be without a parent?

The issues raised about in vitro fertilization of human eggs are profound and the views on these issues disparate. Nevertheless, virtually everyone expressed the opinion that only a continuing dialogue between scientists and public would permit a thorough exploration of the issues and an ultimate consensus.

Additional Reading

1. B. G. Brackett, in *The Biology of the Blastocyst*, R. J. Blandau, Ed. (Univ. of Chicago Press, Chicago, 1971), pp. 329-348.
2. R. L. Brinster, in *Growth, Nutrition, and Metabolism of Cells in Culture*, G. H. Rothblat and U. J. Christofalo, Eds. (Academic Press, New York, 1972), vol. 2, pp. 251-286.
3. Y.-C. Hsu, *Dev. Biol.* 33, 403 (1973).
4. D. G. Whittingham, S. P. Leibo, P. Mazur, *Science* 178, 411 (1972).

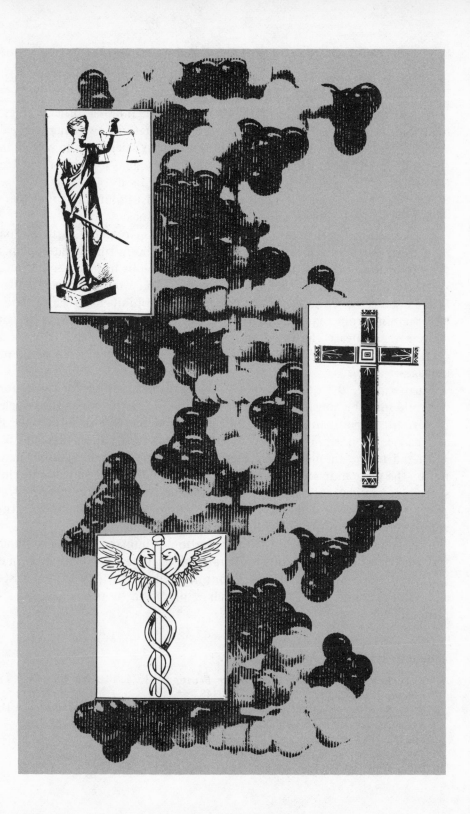

IV SOCIAL, ETHICAL, AND LEGAL PROBLEMS ASSOCIATED WITH ADVANCES IN GENETICS

Views of Scientists, Sociologists, Theologians, and Legal Scholars

In this chapter you will be reading of the diverse and sometimes conflicting views of a wide variety of professionals, all of whom have impeccable qualifications in their respective disciplines. The very fact that these writers look at the problem of genetic and reproductive manipulation from the diverse viewpoints of science and medicine, sociology, theology, and law is almost certain to result in the expression of conflicting views.

Perhaps one of the most encouraging aspects of the articles in this section of the book is the significant number of them authored by scientists who are obviously seriously concerned about the social, ethical, and legal impact of advances in genetics. The last article in the chapter indicates that some scientists are willing to go so far as to propose an embargo on their own research out of concern for the welfare of society.

The first part of this section of the book consists of five brief editorials that document the sometimes differing viewpoints of scientists and physicians with respect to the ethical, social, and legal aspects of genetic and reproductive manipulation. For example, geneticist Richard Lewontin of Harvard University, in an editorial from *BioScience,* suggests that there are ethical issues related to science that are of far greater significance than those associated with genetic counseling and genetic and reproductive engineering. Dr. Lewontin shows that ". . . the activities of science exploit or neglect whole classes of people, taking advantage of their power-lessness."

Two editorials from *Science* present a more conservative picture of the prospects of genetic engineering than one encounters in the popular press. The editorials note some of the dangers associated with arousing the hopes and fears of the lay public with respect to genetic engineering. Certainly, as *Science* editor Philip H. Abelson notes, one of the most unfortunate consequences of this fear would be the placing of harmful restrictions on basic scientific research. Similarly, after expressing their reservations about proposed gene therapy techniques and after expressing their reservations about proposed gene therapy techniques and after noting the limitations and hazards of these techniques, Drs. M. S. Fox of Massa-chusetts Institute of Technology and J. W. Littlefield of Harvard Medical School, stress the need for more public support of basic research in human biology.

The late Dr. Eugene Rabinowitch of the State University of New York at Albany, in an editorial from *BioScience,* posed one of the major ethical questions associated with genetic engineering proposals: "Who will be the decision-makers?" Who will control the application of gene therapy techniques? How will they arrive at their decisions? Who will formulate the rules of the game? Who will play God? Dr. Rabinowitch suggested that the solution to these problems lies in education that stresses ". . . the impact of science on fundmental aspects of human existence."

The final editorial, by Dr. Margery W. Shaw, Director of the Medical Genetics Center at the Unversity of Texas at Houston, treats some of the moral, ethical, and philosophical issues created for the genetic counselor by new genetic discoveries. The genetic counselor works at the grassroots level where counselees are individually advised with respect to procreative problems. Consequently, the genetic counselor sees exactly what ethical and moral problems are created for individuals and families by progress in genetics. Dr. Shaw notes that differing opinions exist among professionals relative to many of these problems, and she argues for preserving diversity of opinion and thus the diversity of people.

The second section of this chapter summarizes reports of groups of diverse specialists with respect to the genetic manipulation question. The first of these articles by Walter G. Peter, III, managing editor of *BioScience,* is a report on an international conference that dealt with ethical perspectives in the use of genetic knowledge. Some of the writers whose articles appear in this book were conference participants. By briefly quoting from the addresses or comments of the participants, Mr. Peter provides a synopsis of some of the conference highlights. One of the best arguments for enhancing the scientific literacy of the general public with respect to advances in genetics is Mr. Peter's own comment, ". . . the uses to which genetic knowledge will be put will not be decided by those who best understand the 'new biology,' but by those who understand it the least, the public or their appointed and elected representatives."

The second article documenting the position of a group is a report from the Research Group on Ethical, Social and Legal Issues in Genetic Counseling and Genetic Engineering of the institute of Society, Ethics and Life Sciences at Hastings-on-Hudson, New York. Dr. Marc Lappé of the Institute served as program director for the report that was signed by 20 notable scientists, theologians, sociologists, and legal scholars. Because asymptomatic heterozygotes for such genetic diseases as sickle cell anemia and Tay-Sachs' disease can be identified in massive screening programs, the need for wise counseling procedures and confidentiality has been increasingly recognized. An especially significant component of the report consists of 11 specific principles designed to serve as guidelines for the operation of a genetic screening program. These guidelines are intended to protect the individuals and families screened and to lessen the risk of misuse of the data obtained in genetic screening programs.

At the present time the technologies for manipulations such as cloning and gene transformation have not been developed to the point where they can be applied to the engineering of humans. We have a period of grace in which to clarify our thinking on the social, ethical, and legal problems that would be created by the application of these procedures to human beings. In contrast, the problems resulting from genetic screening and genetic counseling are with us now. These procedures are being used now, are certainly going to be more widely used in the future, and could become mandatory by legislative action in some cases. The ethical issues associated with these procedures must be faced now. A portion of James R. Sorenson's "Social Aspects of Applied Human Genetics" continues the discussion of the implications of genetic screening and counseling. Dr. Sorenson, Associate Professor in the Department of Socio-Medical Studies of the Boston University School of Medicine, discusses such topics as the nature and scope of genetic counseling, the role of the genetic counselor, and the nature of and basis for reproductive decisions that must be made after genetic counseling.

Dr. Paul Ramsey, Professor of Christian Ethics at Princeton University, is seriously concerned with the bioethical problems created by advances in genetics (See Ramsey's *Fabricated Man,* cited in the bibliography of this book). In May, 1972 Dr. Ramsey and Dr. Paul A. Freund (whose paper follows) presented papers at a "Symposium on Genetic Engineering: Man's Responsibility for His Future" at the Worcester Polytechnic Institute in Massachusetts. In his presentation, Dr. Ramsey emphasized that he sees no major moral or ethical objections to "genetic surgery"—the correction of genetic defects in individual patients. On the other hand, he clearly opposes on moral grounds "germinal engineering"—attempts to prevent the transmission of deleterious genes, the engineering of the child-to-be. Ramsey argues that humans need to give as much consideration to the protection of their own natural biological futures as they are currently giving to the futures of endangered species of animals.

In "Xeroxing Human Beings," Dr. Paul A. Freund, Professor of Law at Harvard University, examines moral and legal implications of advances in genetics. As the title of his paper indicates, Dr. Freund stresses particular issues raised by the prospect of cloning, but he also discusses negative eugenic practices (e.g., abortion, infanticide) and the nature of the moral standards that govern our decisions on such matters. In this latter connection Freund discusses the principle of equal protection of the laws and the political check embodied in this principle.

Continuing with an examination of the ethical and legal implications of genetic engineering is an article from *Saturday Review* by the junior United States senator from California, John V. Tunney, and his legislative assistant, lawyer Meldon E. Levine. These writers cite 10 general considerations that will necessitate making ethical evaluations and supporting our action (or inaction) by moral arguments. Tunney and Levine note that advances in genetics and their application to humans are related to constitutional, statutory, and common law. Finally, Tunney and Levine see a need for considering these problems now rather than at some future date when ". . . we might face irreversible trends not only in genetics but also in political freedoms."

With respect to the application of genetic discoveries to man, biology professor Werner G. Heim of Colorado College differentiates between the role of the biologist and technologist on one hand, and the role of the social scientist, legal scholar, theologian, and humanist on the other. Professor Heim suggests that social scientists, lawyers, humanists, and theologians have the responsibility for assisting society in deciding "...whether, how, and under what conditions the work of [biologists and technologists] ought to be applied." At the same time the biologist should not be absolved of all responsibility for the social, ethical, and legal implications of his discoveries. In fact, Dr. Heim clearly enumerates what he perceives these responsibilities to be. Heim does suggest, however, that the biologist lacks the expertise of the humanist and social scientist in considering the ethical, legal, and social significance of the application of recent genetic discoveries to mankind.

The final paper in this section of the book—a news report by *Science* staff writer Nicholas Wade—reveals that some scientists are so sensitive to the ethical implications of their own research that they are willing to temporarily curtail that research in order to reduce potential health hazards to mankind. The decision by the scientists to call for an embargo on gene insertion studies in bacteria was announced in mid-July, 1974, and was often erroneously reported or interpreted in the popular press. Mr. Wade's account in *Science* is an accurate assessment of what the scientists proposed and of the possible significance of the embargo for human genetic engineering.

16
Science and Ethics

RICHARD C. LEWONTIN

As science, and especially biological science, has gained ever-increasing power to alter our species' relationship to nature and the relationship of one human being to another, there has been a growing preoccupation on the part of scientists with what are conceived of as the "ethical" problems that arise from that power. This preoccupation, however, has been confined almost exclusively to individual ethical decisions, related to problems of individual choice and freedom. Thus, we are told that if we can diagnose genetical disorders *in utero,* a serious ethical problem arises in the decision to deprive a possibly debilitated foetus of its right of life. In a more futuristic vein, we are asked to ponder the weighty ethical problem that arises if we could manipulate an individual's genes at birth or before, since we would be making a decision about a person's biological nature and the biological nature of future generations. Or, again, we are reminded of the serious ethical issues that face the clinical investigator who must decide whether to use an untested and unsure treatment on a seriously ill patient.

What is seldom realized is that the preoccupation with individual moral issues such as these, however serious they may be, is the result of a class bias peculiar to scientists, academics, and other middle class persons. For such privileged persons for whom personal freedom and choice are taken

Reprinted from *BioScience* Vol. 21, p. 799, August 1, 1971 by permission of the American Institute of Biological Sciences.

for granted, these individual interferences with liberty and destiny seem fraught with significance. But they are utterly blind to the fact that large groups of human beings are victims, by *socially determined necessity,* of scientific decisions and research priorities. Why is it that most blood donors in America are either the social outcasts of the cities, or persons who cannot afford to pay for the blood needed by their hospitalized relatives, and so must pay with their own life's blood? Why is it that clinical experimentation is carried out on prisoners, conscientious objectors, the indigent institutionalized, and others who can be induced to trade their bodies for the promise of freedom or a few precious and scarce privileges? Why is it that South Africa's most famous heart transplant had as its recipient a white well-to-do dentist and as its donor a black? How much does a heart transplant cost, anyway, and can someone on welfare get one? What is the meaning of the fact that large sums are newly put into "genetical engineering" and other elegant and sophisticated genetical research, but that black Americans are not routinely screened at birth for the inherited deficiency in the activity of the enzyme glucose-6-phosphate dehydrogenase, which is a potentially lethal disorder present in 14% of black men and 2% of black women? Who pays for the medical research funded by the NIH, and who are the chief beneficiaries?

The questions go on and on, but they seem not to concern our moral philosophers. While the biologist indulges his moral fancies in the ethics of amniocentesis or the problem of whether an extra Y chromosome may foreshadow a life of crime, he turns his back on the fact that the activities of science exploit or neglect whole classes of people, taking advantage of their powerlessness.

The chief and overwhelming ethical and moral issue facing us is that the organization of our political economy guarantees that a large fraction of human beings will be the victims of the omissions and commissions of science because they lack the material wealth and the social power to control their own lives. If biologists are really concerned with the ethics of their activities, they should turn their attention to this social question, and they must have the courage to follow wherever that examination leads.

17
Anxiety about Genetic Engineering

PHILIP H. ABELSON

Science and technology have provided society with innumerable options and the ability to influence evolution. Optimists see in this a great future, with enhancement of the quality of life and of the dignity of mankind. Pessimists see quite a different picture, and at present they appear to be the more numerous and influential. The average citizen, when he thinks about it, is uncomfortable with the necessity of judging complex issues without adequate facts or background. He also feels relatively powerless to affect the outcome.

In spite of the widespread feeling of ineffectiveness, some people have had very great influence and, collectively, the electorate is having profound effects. Public concern about misuse of technology is leading to measures designed to cope with such present abuses as pollution. Technology can be geared to ameliorate part of the disagreeable conditions, and constructive steps are now being taken.

Some of the difficulties created by science and technology are not so close to solution, particularly in biology and medicine. Advances in these fields have led to great benefits and also to puzzling problems, including some for which our present ethical concepts do not prepare us. More technology alone is hardly likely to provide satisfactory answers to the population explosion. Especially disturbing are aspects of the measures

Science, 1971, Vol. 173, Issue 3994 p. 285. Copyright 1971 by the American Association for the Advancement of Science.

taken to prolong life in the very sick and very old. Death of a loved one was bad enough when it was in the hands of God; now it is often a much more distressing experience. Since every individual must participate in birth and death, he cannot escape some thought about the associated problems that science has created; in general, he is not grateful for the necessity to face such issues.

During the last several years, the public has repeatedly been warned that science is creating additional problems through raising the possibility of test tube babies and "genetic engineering." The response of the public has been negative, with some calling for a halt to research in molecular biology. In truth, the dire predictions of the potentialities of new science have outrun the accomplishments, and the predictors have assumed that society will exercise negatively the options that are provided. Speculation about test tube babies is based on a modest accomplishment—that is, fertilizing a human egg in vitro and keeping it alive for a week or so. For many years, biologists have been fertilizing eggs of countless species in vitro. Talk of genetic engineering received impetus from the isolation of an operon, a specific piece of DNA. This accomplishment is about as meaningful as the isolation of other components of the living system. Biochemists are experts at taking life apart, and they can reassemble some subsystems. The total system, however, is orders of magnitude more complex than anything they have put together. Even if biochemists achieve a capacity for genetic engineering, it is unlikely that their tools will match the tools that are already available. For example, artificial insemination is widely used to improve livestock. If some future ruling clique decided to engage in human genetic improvement, they would be more likely to adopt this technique and to employ their own semen than to use material concocted in the laboratory.

Talk of the dire social implications of laboratory-related genetic engineering is premature and unrealistic. It disturbs the public unnecessarily and could lead to harmful restrictions on all scientific research.

18
Reservations Concerning Gene Therapy

M.S. FOX AND J.W. LITTLEFIELD

The attention recently given the prospects of gene therapy requires a realistic appraisal of the potential as well as a sober consideration of the liabilities of this therapeutic approach.

There is no doubt that the development of techniques for transfer of genes and chromosomes in laboratory studies of mammlian cells will provide a powerful research tool toward comprehension of both normal and abnormal cellular processes and will ultimately provide a rationale for the treatment of many human diseases. Gene therapy, however, involves direct application of this technology to individuals suffering from genetic disease; Possibilities under discussion include: introduction of DNA or of chromosomes either directly or by somatic cell fusion; transfer of genetic material from one host to another by virus-like particles containing DNA of the host cell; infection with active or inactive virus containing genes that can determine some particular biochemical function; or infection with a viral nucleic acid to which some cellular gene has been coupled.

Although the number of newborns suffering from disorders that can be described as genetic is very large, only a small fraction of these disorders would even in principle be amenable to intervention by any of these techniques. Neither genetically dominant disorders, nor multigenic traits, nor disorders resulting from extra chromosomes could be alleviated. The

Science, 1971, Vol. 173, Issue 3993, p. 195. Copyright 1971 by the American Association for the Advancement of Science.

major remaining class is that of the recessive "inborn errors of metabolism." These occur with a collective frequency of about 1 per 1000 individuals and include, conservatively, between 100 and 1000 different disorders. Gene therapy would be likely to involve the isolation of somatic cells from a diseased individual, the alteration of their genetic endowment in vitro, and their replacement in the individual. For example, it seems unlikely that sickle-cell anemia would be relieved if a few percent of the blood-forming cells were replaced by cells capable of producing normal hemoglobin, or that the consequences of phenylketonuria would be relieved by the presence of a few somatic cells capable of converting phenylalanine to tyrosine. On the whole, it does not seem probable that more than a small fraction of the inborn errors could be helped by these techniques, and, with new developments in the understanding of the immune response, these disorders will probably be treated more easily and effectively by tissue transplantation or some sort of enzyme therapy.

Furthermore, there are certainly hazards, both known and unknown, that accompany the presently conceived strategies. Many of the procedures are likely to be mutagenic, and who can guess how many dominant effects, visible only in the whole individual, might appear? Most of the viruses under consideration as vectors are tumor-producing. Even the fractionated virus-like particles containing cell DNA are certain to include some particles containing viral DNA. Damaging alterations of regulatory processes and even uncontrolled tumor-like growth could easily be the consequences of introducing additional chromosomes or a host of viral genes.

The promises offered by the proponents of gene therapy largely ignore its limitations and hazards. To mislead the public in this regard risks another period of disappointment and reaction. We are still primarily in a descriptive phase in our understanding of human genetics, with little, if any, idea of how to intervene safely at any level. Let us not do to ourselves what we have done to our environment. Let us now seek public support for research toward a better understanding of normal and abnormal human biology, rather than promise quick glamorous cures

19
Who Will Be the Decision-Makers?

EUGENE RABINOWITCH

It is commonly said that, in the last hundred years, first chemistry, then physics, have had the greatest impact on human existence on earth; and that now this role is being taken over by biology. It seems unlikely, at present, that the era of biological weapons will succeed that of nuclear weapons, as the latter had succeeded the age of chemical explosives; but biochemical and biophysical understanding of life processes clearly is approaching the stage when man will be able to understand and significantly affect the genetic makeup and behavioral patterns of living organisms—including his own species. Man's control of chemical and physical processes has given him immense capacities for constructive achievements as well as for destruction, the latter culminating in almost unlimited destructive power of nuclear explosives.

Breakthroughs in biology promise to put in human hands revolutionary tools for changing human behavior and for reshaping man's existence on earth, for better or for worse. Creation of nuclear explosives has raised the still unresolved questions: who is to control them and who is to determine whether, and for what purpose, they are to be used? The world society cannot remain truly viable if decisions concerning employment of nuclear explosives are not taken out of the hands of individuals, or groups, motivated mainly by self-centered considerations of their individual or collective "best interests"!

Reprinted from *BioScience* Vol. 21, p. 1149, December 1, 1971 by permission of the American Institute of Biological Sciences.

Even more crucial—and totally unprecedented—will be certain problems raised by powers the biological sciences are likely to put into the hands of individuals and societies. Two examples come to mind. One is control of the aging process. Within a few decades, we may well interfere with it. How will mankind control the applications of this knowledge? Will individuals, or their families, have the right to decide to prolong their lives, or those of their loved ones, at will? Will only the ones who can afford the life—prolonging treatment be treated? Or will society insist on giving everybody the same access to it? Or will the state select those worthy of an extra span of life? What will happen to the present-day retirement rules if men and women retain full physical and mental powers for 10, 20, or 50 additional years? What will happen to youth's legitimate desire to take over leading positions in professional life and in the political establishment? Who will determine the answers to these questions, and on the basis of what criteria?

A second example is genetic engineering. We are just witnessing first clumsy steps toward control of certain genes by the roundabout way of infecting cells with viruses bearing these genes. It may be a long time before we develop direct chemical or physical tools to eliminate selectively "undesirable" genes or introduce, or reinforce, other "desirable" ones. But if and when such capacities become a reality, who will control their application, and on what basis? As long as genetic treatment will be used only to eliminate genetic defects, its applications may raise no major ethical, legal, or social problems; but if its potentialities extend to predetermination of physical or behavioral characteristics of the offspring—not only its sex but also its height, physical characteristics, intellectual and artistic tendencies and capacities—who will be allowed to play God, and who will formulate the rules of the game?

These are just two examples of challenges with which biology may well confront mankind—individuals as well as organized societies—in the not too distant future; and we better see to it that the generations succeeding us will be given adequate intellectual and moral foundations to confront these problems wisely, and find adequate answers to them. For this, "education for citizenship"—and that means, in a democratic society, everybody's education—will have to provide not only adequate knowledge of science in general and of biology in particular but, above all, understanding of the impact of science on fundamental aspects of human existence.

This, it seems to me, is the central problem of education in the coming decades. Unfortunately, not much is being done about it.

20
Genetic Counseling

MARGERY S. SHAW

The avalanche of new genetic discoveries and revolutionary techniques (such as amniocentesis, fetoscopy, and fetal biopsy) has introduced into our social consciousness an awareness of moral, ethical, and philosophical concepts at the grass-roots level that were formerly left to the ivory-tower scholars. Suddenly we are faced with choices previously left to fate: Should a genetically defective child be aborted? Should parents be allowed to select the sex of their offspring? Do parents have an inalienable right to knowingly produce defective children?

These and related questions are receiving increasing attention in professional journals and in articles for the layman. The bioethicists promulgate heated arguments, convincing me that there are no right or wrong answers to the questions posed. Most of the recent articles on genetic counseling have pointed out the complex issues involved, the difficulties in communication, and the ethical dilemma posed. Many authors have taken a stand on the "best" way to counsel. Each reflects a different point of view.

My point of view is that we need to preserve these different points of view. As a geneticist, I cherish human variability. Our collective strength lies in our pluralistic attitudes. Freedom of speech preserves our right to shout our convictions and to try to persuade others. But the power to

Science 1974, Vol. 184, Issue 4138, p. 751. Copyright 1974 by the American Association for the Advancement of Science.

make rational decisions preserves our ability to resist coercion from others. There is no opportunity for rational choice, however, when a counselor does not disclose this variety of viewpoints which enables the counselee to arrive at an independent, autonomous decision.

I am not afraid that genetic screening will lead to genocide, nor that abortion will lead to infanticide, as many have warned. If we need checks on our behavior the law will provide them.

The relationship of the genetic counselor to the counselee is a very personal one. It involves a discussion of procreation decisions. Should the counselor describe only the genetic risks, inform the counselee of the range of decisions available, or advise the counselee on the action to be taken? Different counselors do different things. The counselor has freedom to persuade, according to his personal convictions, but he does not have freedom to coerce, based upon his inherent power in the counseling milieu. He must accept the counselee as the ultimate decision-maker. Different parents have a variety of motives for their ultimate decisions. Thus, the outcome of their deliberations will vary. And we will preserve our genetic heterogeneity!

Professionals in general, and physicians in particular, tend to adopt a paternalistic attitude in dealing with patients or experimental subjects or relatives. But only those who desire "parenting" will blindly follow another's advice. The rest will be influenced to a greater or lesser degree by the prescriptions of the counselor who is directive rather than permissive. Those who would argue that the counselee needs protection from directive counseling are themselves being paternalistic.

As our life-styles have become more individualized, so have our ethical codes. The Supreme Court has recognized this trend in the Roe *v.* Wade decision on elective abortion. The proposed amendment to the Constitution to reverse this decision would narrow our options to make individual choices. I see no immediate need to arrive at a consensus or to make policy decisions. Let us keep our options open and preserve our diversity of opinion. By so doing, we preserve our diversity of people.

21
Ethical Perspectives in the Use of Genetic Knowledge

W. G. PETER, III

"Ours is a time of intense self-doubt, corroding confidence, and crippling resolve; a time of troubled present and ominous future . . . and hence it is not surprising that so great a triumph as man's discovery of the molecular basis of inheritance should provoke fear instead of joy, breed suspicion instead of zest, and spawn the troubled anguish of indecision instead of the proud relief of understanding." With these words, Robert Sinsheimer, California Institute of Technology, expressed the growing anxiety with which scientists and laymen are viewing the increasing potentialities of the "new biology." It is this anxiety which brought 80 internationally renowned scholars to a secluded 4-day conference in the Washington, D.C. suburb of Warrenton, Virginia.

Jointly sponsored by the Institute of Society, Ethics and the Life Sciences and the John E. Fogarty International Center for Advanced Study in Health Sciences (NIH); the subject under discussion was "Ethical Issues in Genetic Counseling and the Use of Genetic Knowledge." It has become routine for observers of like meetings to write in summation that "meaningful dialogue was begun from which it is hoped that fruitful results will follow," or that "the necessary groundwork was laid from which further discussions can now begin." These stock refrains are usually an indication that a meeting was a disaster but the motives were noble. In the nonperjorative sense, both of these expressions accurately summarize

Reprinted from *BioScience* Vol 21, pp. 1133-1137, November 15, 1971 by permission of the American Institute of Biological Sciences.

this conference. Meaningful dialogue *was* begun and necessary ground-work *was* laid for further discussions. It remains to be seen if such a large conference can achieve greater goals.

Although the discussions were quite far-reaching, there seemed to be a constant reminder that the uses to which genetic knowledge will be put will not be decided by those who best understand the "new biology," but by those who understand it the least, the public or their appointed and elected representatives. The jurists and physicians who attended the conference often took the role of Public Advocate, although from different and sometimes conflicting positions, while the geneticists pondered the more elusive questions of theory, probability, and potentiality. It would be unfair, however, not to mention that several "cross-overs" took place whereby unlikely allies found themselves in the same camp. Genetic counseling and recent advances made in the detection of defective fetuses, primarily amniocentesis, brought into focus many ethical and legal questions. For example, geneticist John Littlefield, in his presentation "Prenatal Genetic Diagnosis: Status and Problems," stated that if a couple seeks a prenatal diagnosis of an inherited genetic disease, one presupposes that such a couple will elect therapeutic abortion if such disease is discovered. Although "medically indicated," this presupposition disturbed those who felt that therapeutic abortions could not be considered an "automatic" solution.

As miniaturized kinetic and enzyme assays become available, prenatal diagnosis will become shorter and the possibility of aborting defective fetuses greater. Perhaps, as Littlefield speculates, it may become possible to diagnose *in utero* serious autosomal dominant conditions such as neurofibromatosis, tuberous sclerosis, retinoblastoma, and Huntington's chorea. Ultimately, screening for all chromosomal and metabolic disorders may be done from amniotic fluid cells, thus making routine the monitoring of all pregnancies.

Daniel Callahan, philosopher and director of the Institute of Society Ethics and the Life Sciences, sees a potential danger to the right of self-determination should such mass screening become "routine" and abortion "automatic." He pointed out that until recently there was no "causal logic" discernible in bearing a defective child, therefore no burden of choice fell upon the unfortunate parents. If people are presented with a "freedom of choice" through the advances in genetics, then "giving people freedom of choice is to make them responsible for the choices they make. It is then only a very short step to begin distinguishing between responsible and irresponsible choices, with social pressure beginning to put in an appearance. Thus while, in principle, the parents of a fetus with a detected case of Down's Syndrome are still left free to decide whether to carry it to term, it is not difficult to discern an undercurrent in counseling literature and discussion that would classify such a decision as irresponsible. This is amplified in a subtle way. Abortion is said to be 'medically indicated' in such cases, as if what is essentially an ethical decision has now become nothing but medical." Callahan made two further points along this line. Along with the basic "right to life" issue is the one of financial costs. When one throws the "cost-benefit analysis" into the "genetic equation," we can now put a "price on everyone's head." He further stated, ". . . behind the human horror at genetic defectiveness lurks, one must suppose, an image of the perfect human being. The very language of 'defect,' 'abnormality,' 'disease,' and 'risk' presupposes such an image, a kind of prototype of perfection." This last point poses obvious questions of the notions of perfection and imperfection with all the political, social, and economic overtones.

Callahn did not reject the further development of genetic knowledge or the art of genetic counseling, but questioned the "spirit in which such an effort is taken . . . the kind of philosophical perspectives which lie behind it, and . . . the social context in which it is carried out."

These concerns brought out the fundamental issues of the rights of individuals vis-a-vis the rights of society. On a philosophical basis, Professor Aiken of Brandeis University saw at the heart of this issue the very definitions of "life" and "rights." He viewed the term "human being" as a normative concept based upon a special hierarchy of rights. He defined a "right" as an "expectation" plus "responsibility" or "obligation" in regard to that which is "expected." He defined "man" as simply a species name which represents a certain order of biolocial life. Therefore, Aiken saw the question of "right to life" as "right to life as a human being" and "right to life as man." In the first instance, biological life must be present but, in the second instance, biological life could exist without the qualities of humanness. Thus, "a newborn baby is not as it stands a human being . . . it has the capacity if circumstances are favorable of becoming one." The implications of this stance are interesting in determining just when a newborn achieves the status of human being and when an aged human being, approaching a state of senility, loses his "humanness." Does Professor Aiken accept the possibility of infanticide and/or the selective culling of the species on the basis of rights lost? In fairness, Aiken was not given the time to substantiate fully his stands nor did he have the opportunity to explore these possible extrapolations. He admitted to overemphasis in order to make several contentions in a brief period of time.

The Practical Effects of Genetic Knowledge

The jurists had difficulty in differentiating between the concrete physiological and the abstract meanings of "life." As discussed by Blair L. Sadler in his paper, "The Law and the Unborn Child: A Brief Review of Emerging Problems," the fetus has been accorded the rights of a person in American Courts. For instance, the law provides for recovery by a deformed child if someone negligently hurt the mother while pregnant. Although there are some exceptions, "there is a distinct trend in the property, criminal and tort laws to regard the fetus as a human being from the time of conception and to accord the fetus rights consistent with this recognition." The courts have been steadily reversing anti-abortion laws while incurring inconsistencies with the above assertions. In the Wisconsin case, *Babbitz v. McCann*, a three-judge Federal Court stated: "Upon a balancing of the relevant interests, we hodl that a woman's right to refuse to carry an embryo during the early months of pregnancy may not be invaded by the state without a more compelling public necessity than is reflected in the statute in question. When measured against the claimed rights of the embryo of four months or less, we hold the mother's right transcends that of such an embryo." The operant words in these cases are "embryo" and "fetus" and the legal definition given to "human being."

Another case, *Geitman v. Cosgrove*, was described by Sadler and discussed by two other jurists at the Conference, Alex Capron of Yale and Lord Kilbrandon of the Scottish Law Commission. The case involved an infant who was born with serious birth defects after his mother had contracted German measles during her pregnancy. The infant and the mother both alleged that the defendant physicians were negligent in not informing the mother of the possible effects of German measles upon the infant, then in gestation. The State presented the infant plaintiff's case as follows: "But for the negligence of defendants, he would not have been born to suffer with an impaired body. In other words, he claims that the conduct of the defendants prevented his mother from obtaining an abortion which would have terminated his existence, and that his very life is 'wrongful'." Since tort law is based upon a doctrine of "compensatory damages," the court concluded that it "cannot weigh the value of life with impairments against the nonexistence of life itself."

Lord Kilbrandon supported the court in its decision and objects to those critics who reduce the argument to one where life without defects can be assigned a positive value (+), nonexistence a no-value (0), and life with defects a minus value (−). Kilbrandon states that "when an author attempts to express his diagram in words, he is forced into the use of an indefinite pronoun which conceals a semantic vacuum." "Such analysis," he (the author) says, "assumes that life without defects is to be desired most, but that in certain situations it would be preferable not to exist rather than to endure life incapacitated by severe physical and mental defects." Kilbrandon continues, "But for the words 'it would be preferable not to exist' you substitute as a plaintiff must do, the words 'I would prefer not to exist.' The fallacy is plain; the 'I' which is the subject being *ex hypothesi* nonexistent in one alternative cannot predicate his preference for another alternative. It would be like saying, 'I am glad I do not exist, because if I did, I would be mentally and physically handicapped.' This is non-language."

Alex Capron did not find fault with the decision concerning the infant plaintiff's suit, but he did find that the court was at fault for extending this line of thinking to the mother's and father's complaint. The court rejected the claims of mother and father who wished to be compensated for the emotional problems and expense of raising their deformed child. The court concluded, "The right of their child to live is greater than and precludes their right not to endure emotional and financial injury." It further contended that "even if such alleged damages were cognizable, a claim for them would be precluded by the countervailing public policy supporting the preciousness of human life." Capron took the position that the parents were denied two rights: the right of informed consent and, thusly, the right to self-determination. Ethically and legally, Capron contended that the physicians were at fault for not (1) explaining the effects of rubella upon the fetus, and (2) not explaining the alternatives open to the parents in regard to genetic counseling and possible abortion. The withholding of information by a physician is tantamount to denying a patient of "due process," and thus negligence.

Charles Fried, Professor of Law at Harvard, saw both Kilbrandon and Capron as missing the essential issue of developing a normative concept of human nature. While they delineated the immediate legal questions, they failed to realize that without a resolution of this essential problem, no solutions will be forthcoming. "The problem of rights, rights of individuals against each other, against the collectivity, even against the public good, is a modern problem. The modern liberal or individualistic concept of personality presupposes a concept of rights which the individual can claim even against the socially defined good. This concept of right is heavily involved in the issues which concern us here. Even as to the special question of genetic counseling, we ask whether a person has the right to certain kinds of information, thy right to have his case treated confidentially, the right to an abortion, or conversely the right to marry and have children, and from a different angle again the right to be born, once conceived. In other areas of the 'new biology' there has been talk of the right to die, the right to be genetically unique, the right to be the product of normal sexual intercourse and pregnancy. . . . What we are far from having is a comprehensive theory of rights, much less a unified theory which shows how rights and the notion of the public good are functions of some overall scheme." What is "good," what do we "want," how shall we "choose" the wants and values of future generations?

As pointed out by Sinsheimer, and made obvious by the above discussion, our empirical advances in the field of genetics have far outstripped our ethical resolves. Cloning, extra-uterine gestation, and in vitro fertilization are causing scholars in all disciplines to search for new definitions, or reaffirmations of old definitions concerning "Life," "Rights," and "Genetic Defect."

Of immediate concern, as indicated by the Gleitman case, and of great importance to the physicians present, were two pressing problems: deciding upon whom the burden of disease is greatest and how to reconcile professional ethical norms when in conflict with the ethics of laymen. In the first instance, the burden might be carried exclusively by the parents of a defective child when that child is without pain or stress. In the case of a mongoloid child, the parents may bear enormous emotional and psychological stress, yet the child may be happy and, within its limits, well adjusted. Jerome LeJeune, geneticist from the Institut de Progenese, Paris, France, questioned how a scientific judgment could be made that would differentiate this kind of child from one that "should live." If suffering is to be a criterion, how do we assign values to the least possible suffering in this case. The child does not suffer at all as far as we know; it is the parents who suffer. How do you quantify suffering, how do you decide what is unbearable? In the second instance, several philosophers and ethicists were quick to point out to the physicians that if indeed a difference existed between their personal ethical standards and those of their profession that they were leading an unconscionable double life. Considerable disagreement was generated, however, concerning the disclosure of information by physician to patient.

A physician is bound by oath not to "do harm." A patient has the right to self-determination and informed consent. These two binding ethics can easily clash. If the jurists were persuasive in theory, specific examples given by physicians showed the anguish that the rigid enforcement of a principle could produce. One example that brought out the dilemma was the case of a child produced by an extra-marital union unknown to the "parents." Such cases have occurred when genetic counseling has been sought and such illegitimacy is discovered. If the family is happy and well-adjusted, what are the obligations of the counselor? Another case presented involved a patient of one of the attending physicians. The patient inherited a rare anomaly known as testicular femination. Although possessing imperfect testicles in place of ovaries and XY chromosomes, she is quite definitely female in *all* other ways and happily married. The gonads must be removed since there is a high risk of cancer involved in such cases. Should the physician "tell the truth"? If the physician did reveal the truth, what of the possible psychological consequences? Could he be sued for "causing harm" if his patient were to become acutely depressed or if the marriage ended in divorce?

Is Mass Screening Desirable?

Another immediate reality of genetic knowledge is the use of mass screening techniques to identify the carriers of genetic traits for disease. In Baltimore and Washington, D.C., a highly successful screening was undertaken under the direction of Johns Hopkins University physician, Michael Kabak, for a very rare anomaly, Tay Sachs disease. Striking the child of parents who are heterozygous carriers of East European Jewish descent, the disease is 100% lethal within the first 6 years of life. It is neurodegenerating, progressive, and show the symptoms of blindness, total paralysis, and total mental retardation. As far as can be determined, the child suffers no pain. The financial cost of maintaining a Tay Sachs baby is about $40,000 annually. The chances of a couple heterozygous for Tay Sachs of bearing a child with the disease is 1 in 4. It can be detected in a fetus during the early second trimester of pregnancy; therefore, the use of amniocentesis and, in the event of a positive diagnosis, therapeutic abortion offer a viable option to such carriers to safely have normal children each and every time.

Prior to the screening, a carefully planned educational program was undertaken so that the community to be screened was knowledgeable about the disease, the screening procedures, and confidentiality was ensured.

By contrast, there has been widespread pressure to screen for the sickle-cell trait—carried only by our Black population. Robert Murray, Chief of the Medical Genetics Unit of Howard University, sees such screening as playing upon the fears and anxieties of a carrier group for which there is no therapy available. A metabolic disease, sickle-cell anemia does not follow Mendel's law of probability. As Murray put it, "What do you tell a Black man or woman who carries the sickle-cell trait?" Outside of the possible stigma attached to being different, hardly a needed further detriment to a Black person, he saw no value in mass screening for this disease.

Murray further stated that the incidence of sickle cell anemia is so low that it hardly ranks with the other problems facing the Black man. He sees the mass screening of this population as symbolically very important to those who advocate it and, realistically, inconsequential to Blacks. David Eaton, a Unitarian Minister and well-known community leader in Washington, D.C., agreed with Murray and stated unequivocally that he viewed such measures as further evidence of the White man's paternalism. He said that he and most Blacks were fighting for survival by combating malnutrition, drug abuse, communicable disease, and a host of other poverty-related ills of far greater consequence than genetic disease. He queried, "How about trying to do something to save healthy fetuses?"

What Priorities Should Exist?

At the end of the Conference, James V. Neel, University of Michigan, presented a list of priorities in the use of scientific genetic knowledge with the following introduction: ". . . I am sure we would agree that for the next 10-20 years, the impact of failure to apply the new genetic knowledge to human problems is far less than failure to solve such major issues as control of pollution, decay of cities, or a sane energy policy. On the other hand, as I believe has been apparent from previous presentations, the potential of genetic knowledge for the good of man is really very great. Furthermore, there is a symbolism about the use of genetic knowledge to influence the genetic composition of the next generation which readily captures the public imagination. Herein lies one of our problems in setting priorities. Right now we are something of a glamour field. . . . We do have important contributions to make to human well-being. However, we have caught the public imagination, and it's going to take a great deal of intellectual honesty and sobriety not to take advantage of this fact."

Neel then presented the following list of priorities: (1) genetic counseling and its logical extension, prenatal diagnosis; (2) genetic screening; (3) genetic repair, either somatic or germinal; (4) prevention of mutation; (5) amelioration of genotype expression; (6) germinal choice; and (7) stabilization of the gene pool. The advances in these areas could be measured against the following criteria: (1) the reduction of the proportion of persons with genetic disease, an objective with which we most readily associate genetic counseling, prenatal diagnosis, and, perhaps, genetic surgery; (2) the creation of genetically superior individuals, by artificial insemination, perhaps cloning; (3) the improvement of the expression of existing genotypes by wide-ranging medical, social, and nutritional methods; (4) the protection of the present gene pool by a world population policy which will at least ensure that as little as possible of what now exists is lost, and damage through exposure to mutagens is minimized; (5) a minimum of incalculable genetic and somatic risks.

Neel asked "let the record show that before we began our discussion of the difficult issue of setting priorities in this field, we recognized that some might consider this discussion both presumptuous and naive. Our discussion should deserve neither of these adjectives as long as we recognize this as an exercise in social and scientific judgment, approached with the necessary humility and objectivity."

There are many questions that were not examined. It is a well-known biological principle that heterogenity and variety are insurors of long species survival. Should eugenics be used to maximize specific limited characteristics, such as might occur with cloning, would this not substantially lower our survival chances? As population control through the limitation of progeny becomes reality, prospective parents will be more concerned about the quality of their children. Will they demand the "right" of genetic counseling? How will we decide which defects constitute reason for abortion and which do not? Will future genetic knowledge alter our understanding of the term "disease"? Will what is now considered a mild or inconsequential anomaly become an undesired defect? Who will make these determinations?

These are but a few examples of issues that will be forthcoming if genetic knowledge continues to proceed at its present pace. There is some question in the minds of geneticists as to whether such research should even proceed or, if it does, whether it should ever be made public. In the words of Sinsheimer, "Much of the despair of our time stems from the realization that—at last—after all the toil and all the inventions and all the savagery and all the genius—the enemy is 'us.' Our deepest problems are now 'man made.' Their origin lies in the inherent corrosion of imperfect man. Should we arm this creature with vast new powers?"

22
Ethical and Social Issues in Screening for Genetic Disease

A report from the Research Group on Ethical, Social and
Legal Issues in Genetic Counseling and Genetic Engineering
of the Institute of Society, Ethics and the Life Sciences

MARC LAPPÉ, PH. D., PROGRAM DIRECTOR; AND JAMES M. GUSTAFSON, PH. D.; AND RICHARD ROBLIN, PH. D., CO-CHAIRMEN

Abstract. The potential advent of widespread genetic screening raises new and often unanticipated ethical, psychologic and sociomedical problems for which physicians and
the public may be unprepared. To focus attention on the
problems of stigmatization, confidentiality, and breaches of
individual rights to privacy and freedom of choice in
childbearing, we have proposed a set of principles for
guiding the operation of genetic screening programs. The
main principles emphasized include the need for well
planned program objectives, involvement of the communities immediately affected by screening, provision of equal
access, adequate testing procedures, absence of compulsion,
a well defined procedure for obtaining informed consent,
safeguards for protecting subjects, open access of communities and individuals to program policies, provision of
counseling services, an understanding of the relation of
screening to realizable or potential therapies, and well
formulated procedures for protecting the rights of individual and family privacy.

Printed with permission from *The New England Journal of Medicine* Vol. 286, pp. 1129-1132,
May 25, 1972.

In recent months a number of large-scale genetic screening programs for sickle-cell trait and sickle-cell anemia, and at least one for the carrier state in Tay-Sachs disease, have been initiated. Further proliferation of genetic screening programs for these and other genetic diseases seems likely, and in some cases participation in these programs may be made compulsory by statute.* Since screening programs acquire genetic information from large numbers of normal and asymptomatic (e.g., carrier state) individuals and families, often after only brief medical contact, their operation generally falls outside the usual patient-initiated doctor-patient relation. As a result, traditional applications of ethical guidelines for confidentiality and individual physician responsibility are uncertain in mass screening programs. Thus, we believe it important that attempts be made now to clarify some eithical, social and legal questions concerning the establishment and operation of such programs. Although we recognize that there are deep divisions regarding the morality of abortion and that certain views would question prenatal diagnosis so far as it involves abortion, we shall not discuss these issues here. In what follows, we have considered the goals that genetic screening programs may serve and have described some principles that we believe are essential to their proper operation.

Goals Served By Screening

It is crucial that screening programs be structured on the basis of one or more clearly identified goals and that such goals be formulated well before screening actually begins. We believe it will prove costly in scientific and human terms to omit or defer a careful evaluation of program objectives. Although there are three distinguishable categories of goals that screening programs may serve, we believe the most important goals are those that either contribute to improving the health of persons who suffer from genetic disorders, or allow carriers for a given variant gene to make informed choices regarding reproduction, or move toward alleviating the anxieties of families and communities faced with the prospect of serious genetic disease. The following are representative statements of goals that have been used to justify screening programs.

*Massachusetts approved an act (Chapter 491 of Acts and Resolves, 1971) on July 1, 1971, "requiring the testing of blood for sickle trait or anemia as a prerequisite to school attendance."

The Provision of Benefits to Individuals and Families

Such benefits may arise from enabling couples found by screening to be at risk for transmitting a genetic disease to take genetic information into account in making responsible decisions about having or not having children. This usually is done by providing genetic counseling services and informing couples about the nature of existing alternatives and potential therapies (e.g., sickle-cell screening). Another advantage consists in detecting asymptomatic persons at birth when amelioration of the sequelae of a genetic disease is already possible—e.g., screening for phenylketonuria (PKU). Still another is providing means for couples, found at risk by screening, to have children free from a severe and untreatable genetic disease (e.g., Tay-Sachs screening).

Acquisition of Knowledge about Genetic Disease

Laboratory research and theoretical studies have had a major role in helping to understand fundamental aspects of human genetic diseases. In addition, however, some large-scale screening programs may be needed to determine frequencies of rare diseases and to establish new correlations between genes or groups of genes and disease. In some such screening programs, no therapy may be immediately available for the pathologic condition, although the information derived from them may lead to therapeutic benefits in the future. Research programs aimed primarily at the acquisition of genetic knowledge per se are important. Yet we believe their value is enhanced when they also contribute information that is useful for counseling individuals or for public-health purposes.

Reduction of the Frequency of Apparently Deleterious Genes

Although little is known about the possible beneficial (or detrimental) effects of most deleterious recessive genes in the heterozygous state, the reduction of their frequency would be one way to decrease the occurrence of suffering caused by their homozygous manifestations. Nevertheless, as a goal of screening programs, the means required to approach this objective appear to be both practically and morally unacceptable. Virtually everyone carries a small number of deleterious or lethal recessive genes, and to reduce the frequency of a particular recessive gene to near the level maintained by recurrent mutation, most or all persons heterozygous for that gene would have either to refrain from procreation entirely or to monitor all their offspring in utero and abort not only affected homozygote fetuses but also the larger number of heterozygote carriers for the gene[1-3] However, substantial reduction in the frequency of a recessive disease is possible by prenatal screening and selective abortion, or by counseling persons with the same trait to refrain from marriage or childbearing.[3] Nevertheless, these means of reducing the suffering concomitant to recessive disease raise moral questions of their own.

Principles for the Design and Operation of Screening Programs

Attainable Purpose

Before a program is undertaken, planners should have ascertained through pilot projects and other studies that the program's purposes are attainable. Articulating attainable purposes is necessary if the program is to avoid promising (or seeming to promise) results or benefits that it cannot deliver. It is also desirable to update program design and objectives continually in the light of the program experience and new medical developments. Consideration might also be given to incorporating additional purposes—for example, sickle-cell screening programs might profitably enlarge their scope to include other hemoglobinopathies[4] as well as general screening for anemia.[5]

Community Participation

From the outset program planners should involve the communities affected by screening in formulating program design and objectives, in administering the actual operation of the program, and in reviewing results. This involvement may include the lay, religious and medical communities as in the Baltimore Tay-Sachs program.[6] Considerable effort should be expanded to make program objectives clear to the public, and to encourage participation. Recent articles describing detection programs for Tay-Sachs-disease heterozygotes[6] and for persons with sickle-cell trait or disease[7] have stressed the educational aspect of program design as the crucial component of successful operation. The principal value of community participation is to afford individuals knowledge of the availability and self-determination in the choice of this type of medical service. Educated community involvement is also a means of reducing the potential risk that those identified as genetically variant will be stigmatized or ostracized socially.

Equal Access

Information about screening and screening facilities should be open and available to all. To make testing most useful for certain conditions, priority should be given to informing certain well defined populations in which the condition occurs with definitely greater frequency, such as hemoglobin S in blacks and deficient hexosaminidase A (Tay-Sachs disease) among Ashkenazi Jews.

Adequate Testing Procedures

To avoid the problems that occurred initially in PKU screening,[8] testing procedures should be accurate, should provide maximal information, and should be subject to minimum misinterpretation. For detection of autosomal recessive conditions like sickle-cell anemia, for example, the test used should accurately distinguish between those carrying the trait and those homozygous for the variant gene.[4,9].

Absence of Compulsion

As a general principle, we strongly urge that no screening program have policies that would in any way impose constraints on childbearing by individuals of any specific genetic constitution, or would stigmatize couples who, with full knowledge of the genetic risks, still desire children of their own. It is unjustifiable to promulgate standards for normalcy based on genetic constitution. Consequently, genetic screening programs should be conducted on a voluntary basis. Although vaccination against contagious diseases and premarital blood tests are sometimes made mandatory to protect the public health, there is currently no public-health justification for mandatory screening for the prevention of genetic disease. The conditions being tested for in screening programs are neither "contagious" nor, for the most part, susceptable to treatment at present.[10]

Informed Consent

Screening should be conducted only with the informed consent of those tested or of the parents or legal representatives of minors. We seriously question the rationale of screening preschool minors or preadolescents for sickle-cell disease or trait since there is a substantial danger of stigmatization and little medical value in detecting the carrier state at this age. However, in the light of recent information that sickle-cell crises can potentially be mitigated,[10] a beneficial alternative would be newborn screening that could identify the SS homozygote in early life, and thereby anticipate the problems and complications associated with sickle-cell disease and provide early counseling to the parents.

In addition to obtaining signed consent documents, it is the program director's obligation to assure that knowledgeable consent is obtained from all those screened, to design and implement informational procedures, and to review the consent procedure for its effectiveness. The guidelines available from the Department of Health, Education, and Welfare[11] provide a useful model for formulating such consent procedures.

Protection of Subjects

Since genetic screening is generally undertaken with relatively untried testing procedures[9] and is vitally concerned with the acquisition of new knowledge, it ought properly to be considered a form of "human experimentation." Although most screening entails only minimum physical hazard for the participants, there is a risk of possible psychologic or social injury, and screening programs should consequently be conducted according to the guidelines set forth by HEW for the protection of research subjects.[11]

Access to Information

A screening program should fully and clearly disclose to the community and all persons being screened its policies for informing those screened of the results of the tests performed on them. As a general rule all unambiguous diagnostic results should be made available to the person, his legal representative, or a physician authorized by him. Where full disclosure is not practiced, the burden of justifying nondisclosure lies with those who would withhold information. If an adequate educational program has been offered on the meaning of diagnostic criteria and subjects participate in the screening voluntarily, it may generally be assumed that they are emotionally prepared to accept the information derived from the testing.

Provision of Counseling

Well trained genetic counselors should be readily available to provide adequate assistance (including repeated counseling sessions if necessary) for persons identified as heterozygotes or more rarely homozygotes by the screening program. As a general rule, counseling should be nondirective, with an emphasis on informing the client and not making decisions for him.[12] The need for defining appropriate qualifications for genetic counselors in the context of screening programs and for providing adequate numbers of trained counselors remains an urgent one. It is the ongoing responsibility of the program directors to evaluate the effectiveness of their program by follow-up surveys of their counseling services. This may include steps (taken with the prior understanding and approval of the subjects screened) to determine how well the information about genetic status has been understood and how it has affected the participants' lives.

Understandable Relation to Therapy

As part of the educational process that precedes the actual testing program, the nature and cost of available therapies or maintenance programs for affected offspring, combined with an understandable description of their possible benefits and risks, should be given to all persons to be screened. We believe this is one of the items of information that subjects need in deciding whether or not to participate in the program. In addition, acceptance of research therapy should not be a precondition for participation in screening, nor should acceptance of screening be construed as tacit acceptance of such therapy. Both those doing the testing and those doing the counseling ought to keep abreast of existing and imminent developments in diagnosis and therapy[10,13-35] that the goals of the program and information offered to those being screened will be consistent with the therapeutic options available.

Protection of Right of Privacy

Well formulated procedures should be set up in advance of actual screening to protect the rights of privacy of individuals and their families. We note that the majority of states do not have statutes that recognize the confidentiality of public-health information or are even minimally adequate to protect individual privacy.[16] Researchers therefore have a particularly strong obligation to protect screening information. Consequently, we favor policies of informing only the person to be screened or, with his permission, a designated physician or medical facility, of having records kept in code, of prohibiting storage of noncoded information in data banks where telephone computer access is possible and of limiting private and public access only to anonymous data to be used for statistical purposes.

Conclusions

Even if the above guidelines are followed, some risk will remain that the information derived from genetic screening will be misused. Such misuse or misinterpretation must be seen as one of the principal potentially deleterious consequences of screening programs. Several medical researchers have recently cautioned their colleagues of the potential for misinterpretation of the clinical meaning of sickle "trait" and "disease."[5] We are concerned about the dangers of societal misinterpretation of similar conditions and the possibility of widespread and undesirable labeling of individuals on a genetic basis. For instance, the lay public may incorrectly conclude that persons with sickle trait are seriously handicapped in their ability to function effectively in society. Moreover, protecting the confidentiality of test results will not shield all such subjects from a felt sense of stigmatization nor from personal anxieties stemming from their own misinterpretation of their carrier status. Extreme caution should therefore be exercised before steps that lend themselves to stigmatization are taken — for example, stigmatization can arise from recommending restrictions on young children's physical activities under normal conditions because of sickle-cell trait, or from denying life-insurance coverage to adult trait carriers, neither of which are currently medically indicated. In view of such collateral risks of screening, it is essential that each program's periodic review include careful consideration of the social and psychologic ramifications of its operation.

References

1. Crow JF: Population perspective, Ethical Issues in Genetic Counseling and the Use of Genetic Knowledge. Edited by P Condliffe, D Callahan, B Hilton, et al. New York, Plenum Press (in press)
2. Morton NE: Population genetics and disease control. Soc Biol 18: 243-251, 1971
3. Motulsky AG, Frazier GR, Felsenstein J: Public health and longterm genetic implications of intrauterine diagnosis and selective abortion, Intrauterine Diagnosis (Birth Defects Original Article Series Vol 7, No 5). Edited by D Bergsma. New York, The National Foundation, 1971, pp 22-32
4. Barnes MG, Komarmy L, Novack AH: A comprehensive screening program for hemoglobinopathies. JAMA 219:701-705, 1972
5. Beutler E, Boggs DR, Heller P, et al: Hazards of indiscriminate screening for sickling. N Engl J Med 285:1485-1486, 1971
6. Kaback MM, Zieger RS: The John F. Kennedy Institute Tay Sachs program: practical and ethical issues in an adult genetic screening program, Ethical Issues in Genetic Counseling and the Use of Genetic Knowledge. Edited by P Condliffe, D Callahan, B Hilton, et al. New York, Plenum Press (in press)

7. Nalbandian RM, Nichols BM, Heustis AE, et al: An automated mass screening program for sickle cell disease. JAMA 218:1680-1682, 1971

8. Bessman PS, Swazey JP: PKU: a study of biomedical legislation, Human Aspects of Biomedical Innovation. Edited by E Mendelsohn, JP Swazey, I Traviss. Cambridge, Harvard University Press, 1971, pp 49-76

9. Nalbandian RM, Henry RL, Lusher JM, et al: Sickledex test for hemoglobin S: a critique. JAMA 218:1679-1680, 1971

10. May A, Bellingham AJ, Huehns ER: Effect of cyanate on sickling. Lancet 1:658-661, 1972

11. The Institutional Guide To DHEW Policy on Protection of Human Subjects, Grants Administration Manual, Chapter 1-40. (DHEW Publication No [NIH] 72-102). Washington, DC, Division of Research Grants, Department of Health, Education, and Welfare, 1971

12. Sorenson JR: Social Aspects of Applied Human Genetics (Social Science Frontiers No 3). New York, Russell Sage Foundation, 1971

13. McCurdy PR, Mahmood L: Intravenous urea treatment of the painful crisis of sickle-cell disease: a preliminary report. N Engl J Med 285:992-994, 1971

14. Gillette PN, Manning JM, Cerami A: Increased survival of sickle-cell erythrocytes after treatment *in vitro* with sodium cyanate. Proc Natl Acad Sci USA 68:2791-2793, 1971

15. Hollenberg MD, Kaback MM, Kazazian HH Jr: Adult hemoglobin synthesis by reticulocytes from the human fetus at midtrimester. Science 174:698-702, 1971

16. Schwitzgebel RB: Confidentiality of research information in public health studies. Harv Leg Comment 6:187-197, 169

23
Social Aspects of
Applied Genetics

JAMES R. SORENSON

Recent advances in science and medicine are increasing man's control of the quantity and quality of human populations. The development of highly effective and relatively efficient birth control techniques permits expanded regulation of population size and growth. In addition, advances in medical genetics are making possible growing intervention and manipulation of the genetic quality of human populations.

Progress in medical genetics can alter man's role in evolution. Man is no longer limited to passive acceptance of all inherited characteristics but is rapidly expanding his technological capacity to include the active treatment, selection, and elimination of many individual genetic attributes. These developments pose complex questions of a moral, ethical, political, psychological, or economic nature. For instance, what genetic attributes or constitutions are desirable? Who is to decide? Should genetic anomalies be reduced in a population? Who shall say how or when? As with most technological developments, knowledge that permits increasing intervention and control of the genetic quality of life is accumulating more rapidly than is man's ability to apply this knowledge wisely.

Some of the technological developments that permit control of the quality of human life have not precipitated serious problems. In most Western societies medical research and practice have achieved near

James R. Sorenson, "Social Aspects of Applied Human Genetics" Social Science Frontiers No. 3, pp. 1-14 & 31-33. Russell Sage Foundation. © 1971 Russell Sage Foundation.

complete control of many of the major infectious diseases that have plagued mankind for centuries; such control has met little resistance. A case in point is the development of polio vaccine and the subsequent virtual elimination of poliomyelitis.

The success of such medical advances was dependent on several things, especially prevailing values. For example, the implementation of programs to control infectious disease required a value system in which disease was interpreted as a natural event. If parents felt shame or guilt for the infectious diseases suffered by their children, attempts to treat the disorders and to develop curative methods would have been obstructed. Also, such a value system had to provide for the approval of man's active intervention to control disease. The doctor in Western society has not only received approval for his intervention in disease control but also has achieved a great deal of esteem and prestige from the society he has served.

The idea of treatment and intervention in man's genetic health is not universally accepted today, by either the general public or the medical profession. This is due in part to existing values and beliefs. Parents often experience guilt and shame for genetic disorders in their offspring (Lynch, et al., 1965). More important perhaps, many people including some medical personnel, believe that medical procedures are not capable of correcting or treating any genetic disorders, or that doctors and parents should not attempt genetic intervention (Lynch, 1969).

Through the ages man has interpreted the significance of genetic anomalies in many ways. Sometimes, anomalies were interpreted as the favor of the gods, and at other times they were considered to be portents of divine wrath (Reisman and Matheny, 1969). Today where these beliefs still linger, they limit the use of genetic knowledge by the public and by medical practitioners. Counterbalancing these beliefs and values is a rapidly increasing technological capacity to treat and to select certain genetic conditions. The eventual role of applied human genetics in medical practice and society will reflect the complex intertwining of existing values and beliefs with increasing technical capacity. Neither factor alone will predict man's orientation toward an intervention in his genetic future.

Recent Developments in Clinical Genetics

One of the earliest applications of genetic knowledge in medical practice was genetic counseling, also referred to as "inheritance counseling." This is a form of medical service through which people gain information about their genetic constitution or that of their children. Genetic counseling has had a short history. For much of this history doctors were limited to informing families that they had a disease or disorder that "ran in the family." Such medical counsel was of little practical use to families. With the advent of Mendelian genetics and much later the linkage of selected diseases with specific modes of inheritance, genetic counseling became more precise. Doctors could give parents with an identifiable genetic disorder statements of the risk that any child they had would exhibit the genetic defect. For example, a couple might have had a child who died from cystic fibrosis, a lethal genetic disease of childhood. A counselor making this diagnosis would inform the couple that they faced a risk of approximately 1 in 4 of having another child with cystic fibrosis (Carter, 1969). The couple could then use this information to make a decision about future reproductive behavior.

Genetic counseling, as practiced until recently, was limited primarily to the calculation and delivery of such risk statements. Prospective parents could evaluate the risk in a given case and then act on their decision. If they chose to take a risk, they had to live with the child whether he was normal or not.

Several recent advances are changing the nature of genetic counseling. The development of amniocentesis, a procedure by which fetal chromosomes are analyzed for abnormalities, permits intra-uterine detection of several genetic and most major chromosomal defects (*Fogarty International Center Proceedings,* 1970). Using this technique, parents may choose to abort a fetus with an identifiable defect and to have only healthy children. For example, Down's Syndrome, more commonly known as mongolism, a congenital moderate-to-severe form of mental retardation due to specific chromosomal abnormalities, can be detected in utero with almost total accuracy and little apparent risk to the mother or fetus (*Fogarty International Center Proceedings,* 1970). With this procedure the mother need no longer carry such a fetus to term, if the parents so decide and if their doctor and the state concur. The risk of having such a child can thus be removed. Recent developments also permit the detection of an increasing number of genetic defects in persons who carry a genetic anomaly but do not clinically exhibit the disease. For example, carriers of sickle-cell anemia, a form of anemia due to an abnormal type of hemoglobin in red blood cells, limited almost entirely to black populations, can be identified (Carter, 1969). This procedure permits afflicted couples to be advised of their condition prior to having children. Thus, genetic counseling can be given before the birth of any afflicted children, rather than after, as in the past.

In addition to an expanded technical base, many other factors have increased the potential application of genetics in medical practice. First, the role genetic endowment plays is being delineated in more and more diseases. It apparently plays a minor role in the etiology of many diseases, such as cancer and heart trouble; a moderate role in others, such as diabetes; and a major, if not determinate, role in a third class of disease, such as cystic fibrosis, sickle-cell anemia, and Down's Syndrome (Carter, 1969). Until now, genetic counseling has been limited primarily to diseases falling in the third class, but it can be used to advise individuals who have diseases of the first two classes. In these situations clients can be informed to the risk of occurrence of the disease and can be advised to take precautionary measures, such as dietary and environmental management, to reduce their risk of disease. In addition, increasing research efforts and discoveries in behavioral genetics, while not yet applicable, indicate that genetic counseling might eventually involve the role that inheritance plays in the appearance of selected aspects of psychological functioning and possibly some aspects of social behavior (Lindzey, et al., 1971).

Of the diseases in the population that are major threats as health problems, those with a genetic base are constituting an increasingly larger proportion (Carter, 1969). In controlling infectious diseases man has found that imperfections or peculiarities in his own genetic constitution are becoming health threats. This factor will certainly operate to increase the demand for early and extensive application of medical genetics to human populations by both professional medical groups and the public.

Changes in public attitudes and practices are increasing public demand also for more extensive programs in medical genetics. With the increasingly widespread acceptance of population control, it seems reasonable that parental concern for the health of their smaller families will increase. This could include concern for the genetic health of their children. In addition, the acceptance of birth control practices reflects new values and changed attitudes toward reproductive behavior. Not only can reproduction be planned, but with the new developments mentioned above the health of many children who risked disease can now be reasonably assured.

Finally, there may be considerable economic advantages in developing mass screening programs of people likely to pass on specific genetic diseases to their offspring (Scriver, 1970). Therapeutic abortion of fetuses afflicted with Down's Syndrome or other severely disabling diseases detectable in utero can save families much sorrow, and save both the state and the families from a severe financial burden (Danes, 1970). The development of intra-uterine detection techniques makes such screening programs feasible. The changing moral and legal climate surrounding abortion suggests that the number of therapeutic abortions will increase in the near future.

Increased application of genetics in medicine will not necessarily be automatic, however. The use of genetic knowledge in medical practice has given rise to many controversial issues in the recent past and continues to do so. How large a risk should prospective parents take? Should parents make the decision? Should abortion of a fetus be permitted on the grounds that it will be abnormal? Should carriers of severe genetic defects be forbidden to marry, or to have children? These questions are in every respect—ethically, morally, politically, emotionally, legally—difficult to answer. More questions will arise as knowledge of human heredity advances.

The increasing intervention in and control of genetic disease and the selection or avoidance of certain genetic constitutions have fostered debates within and between many different groups in society. Medical professionals, life scientists, lawyers, ministers, biologists, and philosophers are engaged in discussions concerning the proper use of such knowledge. A prerequisite for intelligent discussion of these issues is information about how genetic knowledge is used today. Which doctors give genetic counseling? Which parents seek counseling? What types of reproductive decisions do people make when faced with genetic disorders? Very little information is available on these topics. If man is to understand and to employ the potential good inherent in medical genetic advances, he must begin by determining by whom and how these advances are used. The social sciences, by providing such information, can make significant contributions to discussions on the use of genetic knowledge.

Human Genetics in Medical Practice: A Review and Analysis

Because the application of human genetics in medical practice is relatively new, few studies explore the psychological and sociological aspects of genetics in medicine. The existing literature is both meager and scattered.

There are numerous medical publications concerning the clinical nature of genetic counseling, which is the major use of medical genetics today (Sorenson, 1971a). This literature includes discussions of problems that genetic counselors face, and the types of services they provide. A small amount of literature reports the general public's knowledge of specific genetic disorders. In addition, there are a few studies of the reproductive decisions made by couples after genetic counseling. There are no detailed studies of the distribution in the United States of medical genetic facilities or their services. Finally, little mention is made of the economic aspects of medical genetics.

Most of the available information has been reported by individual medical geneticists or counselors and, being based on their personal experiences, is of limited use in generalizing about medical genetics. There is thus considerable need for extensive social science research on applied human genetics.

Public Knowledge and Use of Medical Genetics

Extensive sociological literature suggests that knowledge about certain kinds of infectious diseases and action taken to alleviate those diseases is inversely related to social class (Feldman, 1966; Mechanic, 1968). The assumption that knowledge about various genetic disorders would vary according to social class is substantiated by existing literature.

In a national survey Feldman (1966) found considerable variation in public knowledge about various diseases. For example, whereas nearly 70 percent of his sample knew at least one correct sympton of polio, only 48 percent could correctly name a symptom of diabetes, a genetically based metabolic disorder (Feldman, 1966). More specifically, knowledge about diabetes was positively associated with education and income, established indicators of social class. The experiences of many genetic counselors support this observation. Their practice suggests that in the past, and to a large extent today, there has been little utilization of medical genetic facilities by the poor (Juberg, 1966).

The fact that historically the majority of people receiving genetic counseling have been of the middle or upper social classes is a reflection of several factors. First, the middle class, and apparently their doctors as well, have been more informed than other classes about the availability of medical genetic services. Second, since the use of genetic counseling has in the past depended largely on self-referral, the willingness of the middle class and the reluctance of the lower classes to seek needed medical assistance have been important (Mechanic, 1968). Economic constraints, while of some importances in shaping the use of medical services, are not always of primary concern (Myers and Schaffer, 1954). Until recently, most genetic counseling was free, with minimal charges for necessary laboratory work. The factors determining the distribution of services appear to háve been essentially special knowledge of the availability of such services, plus sets of values about and attitudes toward the procreative process that legitimated intervention and control.

The social distribution of health information is itself determined in part by the social sources of that information, that is, the process of knowledge dissemination. Feldman (1966) suggests that for diseases such as cancer and polio, essentially nongenetic diseases, a primary source of information is the mass media. His research, as well as several additional studies (Koos, 1954; Deshaies, 1962), suggests variation across social classes, however, in the relative importance of various information sources for knowledge of infectious disease. Generally, the mass media are basic information sources for the more educated segments of the population, while personal contacts serve as primary information sources for the less educated segments (Feldman, 1966). Doctors play a minor role in educating the public. Their role is apparently limited to informing the patient about the particular disease he has (Feldman, 1966).

Public knowledge about genetic disorders does not appear to be disseminated in the same way as knowledge about nongenetic disorders, especially for the more educated segments of the population. Feldman reported that the most important source of information about diabetes was an afflicted relative or close friend. This suggests that general knowledge about genetically based diseases depends more upon inter-personal channels of communication than does knowledge about infectious diseases, and that individuals are likely to be acquainted with such diseases only if a family member or close friend is afflicted.

The social distribution of knowledge about genetic disease reflects the fact that genetic considerations are of limited interest to most people. Concern for these disorders is generally confined to those afflicted, or to those in their reproductive years who have family or friends suffering from genetic disorders. Experiences of counselors indicate that genetics is of minor concern for most people when they select marriage partners. Requests for premarital counseling constitute a small segment of a counselor's time (Reed, 1963). In those cases the most frequent questions concern consanguinity, not the existence of a genetic problem in one partner or the other. The most frequent requests for counseling come from the parents of a genetically defective child (Reed, 1963). It is at this point that the parents become concerned about genetic health and seek counseling to avoid further defective children or simply to relieve anxiety. Counselors are also often contacted for advice by state and local agencies such as state institutions and adoption agencies. Occasionally, counselors are contacted for assistance by the courts in a paternity suit (Reed, 1963).

Taken together, these observations suggest that in the past and at present medical genetic services are not used by all who could benefit from them, nor are they used at the most opportune time. While medical genetics is inherently a preventive form of medicine, it seems to be used in response to crisis situations—after genetic problems occur, rather than before.

The Nature and Scope of Genetic Counseling

There are many forms of genetic counseling. The people who practice genetic counseling have various educational backgrounds: some are research scientists, others medical doctors (Sorenson, 1971b). Genetic counseling is not yet a medical specialty. Because most physicians lack training in human genetics, counseling is usually limited to those who have specialized knowledge about human genetics, although they may or may not be medical doctors.

In the late 1940's and early 1950's there were perhaps 10 or 12 genetic counselors in the United States (Reed, 1963). Today there are about 200 counseling units (Lynch and Bergsma, 1971). With mounting discoveries and an increasing awareness of the need for counseling, this impressive growth should continue in the near future.

The most common task facing the genetic counselor is the determination of the risk of recurrence of a genetic defect in a family (Reed, 1963). Supported by Mendelian genetics, the notions of dominant and recessive autosomal or x-linked modes of inheritance, the counselor can often provide specific information about recurrence risks (Lynch, 1969). Although many genetic disorders may follow the classical inheritance modes of Mendelian ratios, many do not. In these cases the counselor must rely upon empirical statistics of risk. These statistics have been derived from studying the incidence of a disease in many families (Neel, 1958). From these observations the counselor can provide the clients with approximations of the recurrence risk of a specific disease. When the counselor is equipped with neither Mendelian ratios nor empirical estimates of risk, he can inform his clients only of his ignorance.

In addition to statistics of risk the counselor today can employ a battery of new techniques that permit increased detection of genetic and chromosomal abnormalities in utero. These techniques include amniocentesis, examination of the genetic and chromosomal health of the fetus, and fetoscopy, a technique that enables visual inspection of the fetus (*Fogarty International Center Proceedings,* 1970). Advances such as these are applicable only when a chromosomal or genetic defect has a structural or metabolic effect upon the developing fetus. Many genetic defects cannot yet be detected by these techniques. While amniocentesis appears to be used primarily for the detection of Down's Syndrome, fetoscopy is still in a largely experimental stage. These techniques are relatively expensive and not widely available.

The Role of the Genetic Counselor

Genetic counseling as a form of medical service differs from traditional medical practice. A complex relationship exists between the counselor and the counselee in genetic counseling. While some counselors perceive the relationship in the traditional doctor-patient manner, others perceive it as a counselor-client relationship (Sorenson, 1971b). Because the rights and obligations of both parties of this relationship have not been completely defined, the practice of counseling is probably based primarily on the beliefs, attitudes, and skills of the individual counselor.

Some counselors feel their primary task is simply to inform the counselees of the risk involved. They believe that to try to influence the decisions of the counselees is to go beyond the professional responsibility of the counselor. While a counselor may feel that the risk involved in a specific case is great or that a given disorder should not be transmitted to a new carriers, he should refrain from telling the counselees what they should do. Often, in this situation, the counselor will define his relationship with the counselees as a counselor-client relationship (Reed, 1963). In so doing he is stressing the learning nature of the relationship. The client is there to be informed, not to be healed.

Characteristic of the traditional doctor-patient role is the healing relationship in which the counselee yields discretionary power to the doctor (Bloom, 1963). Some counselors describe their relationship with their counselees in this fashion (Lynch, 1969). The counselor may suggest how patients should use the information he gives them, and his concern for the patient goes beyond a mere statement of statistics of risk. For example, the counselor may discuss in detail the psychological consequences of an abnormal child, or the economic burden of genetic abnormality. He may also point out to the parents the impact on the family of a defective child. Finally, he may discuss with them the problem of dealing with an afflicted dependent when they advance in age and retire (Fletcher, 1971). All of these are important factors in conveying the meaning of genetic risks to parents and undoubtedly have an impact on the parents' decisions about future reproductive behavior.

Because his role falls within the purview of medical service, the counselor is probably approached by most people in the traditional patient manner. This orientation requires the counselor to assume much more discretionary power than most want. The meaning of a risk is complex, as is the meaning of disease or abnormality. While many counselors and counselees may agree that it is good to avoid having mongoloid children, there is less agreement as one moves from physiological abnormalities to such problems as an XYY chromosome anomaly and the hypothesized link with social pathological behavior (Public Health Service, 1970).

The information the counselor gives, as well as the way in which he gives it, is crucial in defining the situation for parents. For example, if the counselor indicates that in the United States about 1 in 50 births will result in a severely abnormal child (Carter, 1969), he can label the risk high or low. If he says that the odds for a *normal* birth are 3 in 4, or conversely the odds for an *abnormal* birth are 1 in 4, the facts are the same but the impact of the information on the parents is likely to be reversed. Because counselors are generally aware of the problems of giving genetic counsel, many are concerned about exerting too much influence on the parents.

The Delivery of Genetic Counseling

The practice of genetic counseling appears to be moving from academic departments to departments in medical schools and in public health centers (Sorenson, 1971b). Today, the typical counselor is more often an M.D. or an M.D.-Ph.D. than the traditional Ph.D. geneticist of the past. These shifts have numerous implications for both the delivery and the scope of genetic counseling.

Whether counseling facilities are located in medical schools or public health clinics makes a difference in the distribution and utilization of counseling. Most doctors in medical schools occupy the dual role of teacher and researcher. Research interests are normally pursued by intensive study of a particular facet of disease. For example, if a doctor-researcher interested in genetics focuses his attention on cystic fibrosis, he will become professionally identified through publications and more informal channels as a specialist in cystic fibrosis. As a consequence, both the dictates of his research and his professional identification will restrict the types of patients he counsels. In addition, if the medical school in which the researcher works does not have a large department or program in human genetics, as is often the case, it will become identified as specializing in counseling on limited genetic topics. Patients with other genetic problems will not be referrd to it and must go elsewhere for counseling.

Genetic counseling is beginning to be practiced in some public health centers (*Symposium on Human Genetics in Public Health,* 1964). With this development, the proportion of lower socio-economic groups receiving counseling will most likely increase. Public health nurses, whose professional duties are primarily concerned with lower-income groups, will become a major force in the dissemination and application of genetic counseling. The public health nurse can provide the family with attention at home, a service seldom available in genetic counseling today (*Symposium on Human Genetics in Public Health,* 1964). Genetic problems can, and often do, have a profound impact upon the social-psychological stability of the family (Zuk, 1959). Family follow-ups and counseling should provide for more effective genetic guidance,

Reproductive Decisions After Counseling

There is limited information on reproductive decisions made by patients after counseling. Certainly many factors are important in determining these decisions: (1) the size of the risk, (2) the severity of the potential abnormality, (3) the social and private attitudes of the parents toward abnormality, (4) the economic capacity of the family to endure the burden of a genetic disease, (5) the genetic health of existing childern, and (6) the type of counseling parents receive. In a recent study, Carter (1966) presented a follow-up of the reproductive behavior of 169 couples referred to a genetic counseling clinic in England. His study revealed that the magnitude of the risk had a large impact on the future reproductive behavior of the couples. Fully two-thirds of the high-risk parents, as against one-fourth of the low-risk group, did not have additional children. (Of course, factors other than the magnitude of the risk affect these reproductive decisions.)

As we have seen, couples who are faced with the prospect of a high risk of a serious defect may decide to take the risk and have a baby. If they decide not to take the risk, however, several options are available to them: (1) they may give up plans for having additional children; (2) depending on the nature of the genetic disorder, they may decide on artificial insemination from a donor; or (3) they may decide to adopt a child. As with the decision whether or not to take a risk, decisions among alternative ways to have a child are influenced by many factors. Unfortunately, there are no published data either on these decisions or on the effects upon the decisions of social class, religious beliefs, legal constraints, and financial conditions. Carter (1966) does indicate that within his sample of high-risk couples who decided not to have any additional children adoption was low. In several cases the wife was sterilized, and in many cases the husband.

With few exceptions, much of the preceding information is based on the personal experiences of genetic counselors. While this information is important, more systematically gathered knowledge is needed. The increasingly important role that genetics will play in medicine in the future requires that we understand how people react to and use genetic knowledge, whether they are professional medical personnel or the public. In addition, the important social, ethical, and legal problems beginning to arise from the development of new medical ideas and technology require solid knowledge of the extent and use of genetic counseling today. Knowledge is being applied and precedents are being set that will shape the use of future discoveries.

References

Babbie, Earl R., *Science and Morality in Medicine*. Berkeley: University of California Press, 1970.

Bailey, Richard M., "Economic and Social Costs of Death," 275-302 in O. B. Brim, H. E. Freeman, S. Levine, and N. A. Scotch, eds., *The Dying Patient*. New York: Russell Sage Foundation, 1970.

Birenbaum, Arnold, "The Mentally Retarded Child in the Home and Family Cycle," *Journal of Health and Social Behavior*, 12 (March), 1971, 55-65.

Black Panther, b, No. 2 (May 22, 1971) and 6, No. 17 (April 10, 1971).

Bloom, Samuel W., *The Doctor and His Patient*. New York: Russell Sage Foundation, 1963.

Carter, Cedric, "Comments on Genetic Counseling," Proceedings of the Third International Congress of Human Genetics, 1966, 97-100.

Carter, Cedric, *An ABC of Medical Genetics*. Boston: Little, Brown, 1969.

Clausen, John A., Morton A. Seidenfeld, and Leila C. Deasy, "Parent Attitude toward Participation of their Children in Polio Vaccine Trials," *American Journal of Public Health* 44 (1954), 1526-1536.

Crane, Diana, *Social Aspects of the Prolongation of Life*, 1, Social Science Frontiers. New York: Russell Sage Foundation, 1969.

Danes, Betty Shannon, "Genetic Counseling," *Medical World News* (November 6, 1970), 35-41.

Deasy, Leila, "Socio-Economic Status and Participation in the Poliomyelitis Vaccine Trails," *American Sociological Review*, 24 (1956), 185-191.

Deshaies, J., "Public Knowledge about Chronic Disease Symptoms," unpublished masters thesis, University of Chicago (1962), cited in Feldman (1966).

Epstein, Charles J., "Medical Genetics: Recent Advances with Legal Implications," *Hastings Law Journal*, 21, No. 1 (November, 1969), 35-49.

Farrow, Michael, and Richard Juberg, "Genetics and Laws 'Prohibiting' Marriage in the United States," *Journal of the A.M.A.*, 209 (July 28, 1969), 534-538.

Feldman, Jacob J., *The Dissemination of Health Information*. Chicago: Aldine, 1966.

Fletcher, John "Parents in Genetic Counseling: The Moral Shape of Decision Making." Paper presented at Conference on Ethical Issues in Genetic Counseling and the Use of Genetic Knowledge, Fogarty International Center, National Institutes of Health, October, 1971.

Fogarty International Center Proceedings, No. 6 (May, 1970) Maureen Harris, ed.

Francoeur, Robert T., *Utopian Motherhood*. Garden City, N.Y.: Doubleday, 1970.

Freidson, Eliot, *Profession of Medicine*. New York: Dodd, Mead, 1970.

Juberg, Richard, "Heredity Counseling," *Nurses Outlook*, 14 (January, 1966), 28-33.

Koos, E. L., *The Health of Regionville*. New York: Columbia University Press, 1954.

Lader, Lawrence, "A National Guide to Legal Abortion," *Ladies Home Journal* (July, 1970), 73.

Lee, Nancy Howell, *The Search for an Abortionist*. Chicago: University of Chicago Press, 1969.

Lindzey, Gardner, John Loehlin, Martin Manosevitz, and Delbert Thiessen, "Behavioral Genetics," *Annual Reveiw of Psychology*, 1971.

Lynch, H. T. *Dynamic Genetic Counseling for Clinicians*. Springfield, Ill.; Charles C. Thomas, 1969.

Lynch, H. T., and D. Bergsma, *Birth Defects Genetic Services*. New York: National Foundation—March of Dimes, 1971.

Lynch, H. T., Robert L. Tips, Anne Krush, and Charles Magnuson, "Family-Centered Genetic Counseling: Role of the Physician and Medical Genetics Clinic," *Nebraska State Medical Journal*, 5, No. 4 (April, 1965), 155-159.

Markle, Gerald E., and Charles B. Nam, "Sex Predetermination: Its Impact on Fertility," *Social Biology*, 18, No. 1 (March, 1971), 73-83.

Mechanic, D., *Medical Sociology: A Selective View*. New York: Free Press, 1968.

Menzel, Herbert, and Elihu Katz, "Social Relations and Innovations in the Medical Profession: The Epidemiology of a New Drug," *Public Opinion Quarterly*, 19 (Winter, 1955-1956), 337-352.

Miller, James R., "Human Genetics in Public Health Research and Programming," *Symposium on Human Genetics in Public Health*, Minneapolis (August 9, 1964).

Myers, Jerome K., and L. Schaffer, "Social Stratification in Psychiatric Practice: A Study of an Outpatient Clinic," *American Sociological Review*, 19 (June, 1954), 307-310.

Neel, James V., "The Meaning of Empiric Risk Figures for Disease or Defect," *Eugenics Quarterly*, 5 (March, 1958), 41-43.

Olshansky, Samuel, "Chronic Sorrow—A Response to Having a Mentally Defective Child," *Social Casework*, 43, No. 4 (April, 1962), 190-193.

Packer, H. L. and R. J. Gampell, "Therapeutic Abortion: A Problem in Law and Medicine, *Stanford Law Review*, 11 (1959), 417.

Public Health Service, "Report on the XYY Chromosome Abnormality," *Public Health Service Publication*, No. 2103 (October, 1970).

Reed, S. C., *Counseling in Medical Genetics*, 2nd ed. Philadephia; Sanders, 1963.

Reismann, Leonard E. and Adam P. Matheney, *Genetics and Counseling in Medical Practice*. St. Louis: C. V. Mosby Company, 1969.

Rosen: Harold, *Abortion in America*. Boston: Beacon Press, 1967.

Rosenfeld, Albert, *The Second Genesis: The Coming Control of Life*. Englewood Cliffs: Prentice Hall, 1969.

Scriver, Charles R., "Screening for Inherited Traits: Perspectives," *Fogarty International Center Proceedings*, No. 6 (May, 1970) Maureen Harris, ed.

Sorenson, James R., "Genetic Counseling Biblography." Unpublished manuscript, Princeton University, 1971a.

Sorenson, James R. "Decision Making in Applied Human Genetics: Individual and Societal Perspectives." Paper presented at Conference on Ethical Issues in Genetic Counseling and the Use of Genetic Knowledge, Fogarty International Center, N. I. H., October, 1971b.

Stevenson, A. C., "The Load of Hereditary Defects in Human Populations," *Radiation Research*, Supplement 1 (1959), 306-325.

Symposium on Human Genetics in Public Health. Minneapolis (August 9-11, 1964).

Szasz, T. S., and M. H. Hollender, "A Contribution to the Philosophy of Medicine: The Basic Models of the Doctor-Patient Relationship," *A.M.A. Archives of Internal Medicine*, 97 (May, 1956), 585.

Vernon, Glenn M., and Jack A. Boadway, "Attitudes toward Artificial Insemination and Some Variables Associated Therewith," *Marriage and Family Living*, 21 (1959), 43-47.

Wilson, Robert N., "The Physician's Hospital Role," 414-415 in *Medical Care: Readings in the Sociology of Medical Institutions*. W. Richard Scott and Edmund H. Volkart, eds., New York: Wiley, 1966.

World Health Organization, *The Teaching of Genetics in the Undergraduate Medical Curriculum and in Postgraduate Medicine*, Technical Report Series, No. 238 (1962).

Wright, Sewell, *The Biological Effects of Atomic Radition*. Washington, D. C.: National Academy of Sciences–National Research Council. 1960.

Zuk, G. H., "The Religion Factor and the Role of Guilt in Parental Acceptance of the Retarded Child," *American Journal of Mental Deficiency*, 64, No. 1 (July, 1959), 139-147.

24
Genetic Engineering

PAUL RAMSEY

Three things are said to evidence the wisdom and greatness of the Chinese people. They invented gunpowder and failed to invent firearms. They invented printing, and didn't think of newspapers. They invented the compass and failed to discover America.

A similar attitude, I believe, should be adopted toward future possible applications of biomedical knowledge. Marshall Nirenberg, the great scientist and modest, troubled man who "cracked" the genetic code, has said that we are apt to learn to move genes around long before we can know it is safe to do so. And Leon Kass, executive secretary of the Committee on the Life Sciences and Social Policy, National Academy of Sciences-National Research Council, has written: "When we lack sufficient wisdom to do, wisdom consists in not doing. Caution, restraint, delay, abstention are what this second-best (and, perhaps, only) wisdom dictates with respect to the technology for human engineering. . . . We must all get used to the idea that biomedical technology makes possible many things we should never do" (Science, Nov. 19,1971). Yet how unlikely it is that we will have any such sense of limits there where we need it most— namely, when technology promises mastery over human genesis, and the alleged perfectability of man's natural endowments.

I might begin by musing over the title of this symposium. It reads:

Reprinted by permission of Science and Public Affairs, the *Bulletin of the Atomic Scientists*. Copyright © 1972 by the Educational Foundation for Nuclear Science.

"Genetic Engineering: Man's Responsibility for His Future." This title is in accord with the most basic, silently operating assumption of the modern age, namely that we should do whatever we can do. The questions remaining are only those of timing and in what manner, yet to be worked out, the technologies made possible by the revolution in biological knowledge are to be made operational.

On its face, who can be opposed to "Man's Responsibility for His Future"? As a reasoned proposition, however, it would have to be shown that future evolution is a chief focus and measure of the meaning of responsibility, that one can get a clear idea of that meaning of moral responsibility, and that genetic engineering does not violate clearer, closer-to-home meanings of responsibility among the generating generations of mankind. In our age, the "can do" takes the place of answers to those searching questions.

"Genetic engineering," of course, is a metaphor. We should make very clear what we are talking about, by adopting a limited referent for that expression. Following the experimental effort, now in process, to change the genes of two little German girls who are suffering from the irreversibly degenerative disease known as argininemia, in the future it may be possible to effect gene-changes in other existing individual human beings (whether conceptus or infant): to alter the gene deficiency that causes phenylketonuria (PKU), or cystic fibrosis, or sickle-cell anemia, or Tay-Sachs disease, or simply to tell a person's pancreas to start making insulin (which would seem a better remedy for diabetes than injections). All such gene changes will be treatments. We should call those procedures genetic therapy, in order to locate them alongside the nongenetic treatments now available for some genetic illnesses. The appropriate metaphor for gene-change treatment is "genetic surgery," not "engineering."

That makes it immediately evident that treating the genes will raise no unusual moral questions, no issues not already present in investigational therapeutic trials, or hazardous or last-ditch surgery of other sorts. To send in a messenger virus to change the genetic code words of an existing patient will raise novel ethical issues if, but only if, such treatments are apt to be erratically inheritable in the next generation or if, but only if, the gene structure of that patient's sperm or ova may be affected in ways that will bring unforeseen and possible deleterious consequences upon that patient's progeny. Concerning gene-change treatment I will only remark that unless future bad consequences or unknown genetic perturbations in a patient's progeny can be foreclosed, diet for the correction of PKU should remain the treatment of choice, while gene-change therapy could still be ventured in the case of Tay-Sachs disease (a fatal disease for which there is today no available remedy).

By "genetic engineering" I mean gene changes targeted upon the sperm and ova (or their precursor cells) of suspected carrier couples with the objective of preventing the transmission of life suffering from serious genetic defect. Since "gamete" is the generic term for ova and sperm and their precursor cells, we might call this "genetic engineering" by "gametic manipulation." Let us simply use common parlance and say "germinal engineering."

I acknowledge that this is a stipulative definition of our topic. I hope it is a persuasive one, so we can keep clear an important distinction between genetic engineering of gametes and the genetic treatment of individual patients. Since only in this way can we make clear the serious moral issues we face in contemplating the possibility of genetically engineering the child-to-be.

It should immediately be evident that, as a proposal, genetic engineering (in the limited meaning I give that expression) is first of all a proposal for preventing the transmission of life suffering from serious genetic defect. Such a procedure would have to be compared with available alternatives in the practice of preventive genetic medicine, would have to be compared with optional uses of our rapidly increasing genetic knowledge in responsible parenthood or responsible non-parenthood. Those alternatives are genetically-conditioned marriage licenses, genetically-motivated voluntary sterilization, using three contraceptives at once, abstinence, or hieing yourself away to an old-fashioned Catholic monastery or nunnery.

What would prompt anyone to adopt one of those seemingly radical alternatives as against the promise of germinal engineering that we can go ahead and have better babies anyway? Or what would prompt me to argue that genetic engineering cannot be the indicated, or a choiceworthy, application of our knowledge of genetics?

Simply because, when measured by the principles of sound ethics or by the canons of medical ethics, germinal engineering would be an immoral experiment on the child-to-be—immoral because it is not consented to by the primary subject; immoral because, when he is not yet, the child suffers no defect which could justify anyone to give such consent on his behalf, or justify a physician in making the risk-filled balancing judgment. Such a judgment warrants investigational trials on actual patients. Rightly ordered concern for the child-to-be compels us to conclude that we ought not to choose for another the possible hazards he must bear in order to suffer our experiments and to be the vehicle of medical "progress" whatever the consequences.

When pressed, researchers in this field cannot deny that there may have to be not a few mishaps, not a few monstrosities discarded, more serious defects induced in place of the one prevented, before they can get to know how to perfect the technology of germinal engineering. The last test done to detect induced damage may itself be injurious or may have to be omitted because it risks greater possible damage; the same for the last intrauterine scanning, the last tap of the amniotic fluid.

The rejoinder that will be forthcoming is as follows. The incidence of probable risk of induced damage can be kept equal to or below the incidence of probable natural defect calculated from the genetic histories or tests of the would-be parents. Moreover, in the course of the necessary testing and scanning of the developing life a lot of other unsuspected defects or accidental damage (as, for example, from the mother's sleeping pills) may be discovered along the way. These can then be treated, or their subject eliminated by genetic abortion.

But the life we are talking about would not have been or been subject to those hazards had he not first been produced by germinal engineering as our means of preventing the transmission of genetic defects. You cannot justify a questionable procedure by appeals that in thought assume it has already been done. By no logic can this new sort of genesis be made to bottom on itself, or be justified by concomitant balancing advantages that already presuppose the experiment (which was in question) that placed the child-to-be risk in of both induced and natural damage. The sperm and egg are in no need of a physician; nor is there any child who has consented to be used at risk to cure his parent's need for a child.

The rejoinder, nevertheless, parses the original proposal. It shows us that the basic idea of genetic engineering or germinal manipulation turns human procreation into the manufacture of our progeny, or rather replaces the transmission of life from life by the categories of manufactory (which of all human activities is most concerned with product design, and with doing anything at all to improve the product, or at least to seem to do so).

Therefore, the proposal to engineer sperm and ova or their precursor cells to prevent the transmission of serious genetic defects contains, in its rationale and in principle, the use of biomedical technology to eliminate minor defects as well as serious ones—positive eugenics no less than negative: the introduction of a gene for blond hair, if desired, no less than genes for greater intelligence or musical ability; the delivery to people of a child with the wanted gender no less than producing from them a child without cystic fibrosis by manipulating their gametes; the production of a dwarf if I need one in the circus I own no less than a potential Horowitz in the family; the mixture of chromosomes from other species with human ova and sperm cultures before they are brought together in a final product on the laboratory assembly line.

These are not just the fancies of the authors of articles in popular magazines today—extrapolations to be denied by serious scientists. Joshua Lederberg of Stanford University begins by speculating on what would be "the effect of dosage of the human twenty-first chromosome on the development of the brain of the mouse or the gorilla"; he ends by contemplating "the introduction of genetic material from other spheres" into the human. And Dr. R. G. Edwards—that "brave" pioneer in the manipulation of embryos at risk—asks whether medical ethics, having been stretched to warrant what he is doing in vitro fertilization and embryo transplantation, can be further stretched to justify "the more remote techniques" for "modifying embryos," such as the production of "chimeras" by adding to a human embryo the precursor cells for organs from other blastocysts, and perhaps from other species.

The speculation that most appeals to me is the idea that I might have been created with a contraption for photosynthesis on my back, so that like a plant I never had to eat, but still had a penchant for philosophy. In any case, the radical displacement of procreation by manufacture happens long before we begin to think of adding to future possible human beings organs and capacities not their own. That begins with the acceptance of genetic engineering as a project. With that acceptance, the fundamental moral argument for all subsequent steps is complete, whether we take those steps or not or whether we judge them to be worth taking.

With genetic engineering by means of gametic manipulation we are already in the world of the Fertilizing Stations and Decanting Rooms of the East London Hatchery in Aldous Huxley's "Brave New World." Huxley's vision of a pharmacological and genetic utopia, we should remember, was an entirely happy society. That means that, defendable step by defendable step, we shall all be happy on the way there; that we are now inclined to be persuaded of the immediate values of each step along the way to the final replacement of human procreation by laboratory manufacture. Otherwise "Brave New World" would not be the happy place it will be. Conversely, if we could not be conditioned to contentment, there would be alarms now, and a wrenching away from the strange notion that opposition to engineering human genesis is somehow retrograde and should be repressed as we "struggle to catch up with what the scientists can do."

C. S. Lewis shares with Aldous Huxley the prescience of having discerned, under the very shadow of Nazism, that genetics rather than the misuse of political power constitutes in our era the greatest danger of "the abolition of man." He wrote in the book of that title in 1942 that we should "not do to minerals and vegetables what modern science threatens to do to man himself." Lewis discerned that the last citadel from which technological applications are apt to be excluded, or where we will discover limits we will agree to defend as we would defend our lives, will prove to be the citadel of man's nature itself. The wounds we have inflicted upon natural objects for lack of a proper sense of the natural environment are becoming clear to us—the lashes and the ecological backlash. Today many are testifying to the spiritual autonomy of natural objects and to arrogance over none, to the scheme of things in which man has his place. But there is as yet no discernible evidence that we are recovering a sense for man as a natural object, too, toward whom a like form of "natural piety" is appropriate.

While the leopard, the great whale, and the forests are to be protected in the ecological ethic of our day by restoring to mankind a proper sense of things, man as a natural being himself is to be given no such protection. There are paremeters of the cheetah's existence that ought not be violated, but none of man's. Other species are to be protected in their natural habitat and in their natural functions, but man is not.

Today the statement of Aristotle is amply verified. It would indeed be odd if a flute player or artisan of any kind, or a carpenter or a cobbler, or the mountain goat and the falcon, "have certain works and courses of action," while "Man as Man has none, but is left by Nature without a work" of his own. It may be that Aristotle's question, What is the work of man, is expressed in language that will seem too functional for the ecological ethics we need in locating man in the creation of which we are a part and toward which we should have a "natural piety." Still his view that all things in nature "have certain works and courses of action" has enough amplitude to be helpful as we search for a sense of man as a natural object too. Procreation, parenthood, is certainly one of those "courses of action" natural to man, which cannot without violation be disassembled and put together again—any more than we have the wisdom or the right impiously to destroy the environment of which we are a part rather than working according to its lineaments, according to the functions or "courses of action" we discover to be the case in the whole assemblage of natural objects.

Playing God

If St. Francis preached to his sisters the birds, spoke of his brother fire, called the turtle doves his little sisters and composed a canticle to his brother the sun (and for this has been proposed as the patron saint of all ecologists), what are we to say of our brother man and of ourselves? If the commandment "to subdue the earth" is a right devilish one or else sorely in need of a loose, or of another, interpretation—an interpretation properly limited by man's vocation to tend the garden of God's creation—are we then to say that man is let loose here with the proper task of disassembling his own "courses of action," making himself and his own species wholly plastic to ingenious scientific interventions and alterations?

So today we have the oddity that men are preparing to play God over the human species while many among us are denying themselves that role over other species in nature. There is a renewed sense of the sacredness of groves, of the fact that air and streams should not be violated. At the same time, there is no abatement of acceptance of the view that human parenthood can be taken apart and reassembled in Oxford, England, New York, and Washington, D.C. And, of course, it follows that thereafter human nature has to be wrought by Predestinators in the Decanting and Conditioning Rooms of the East London Hatchery and in commercial firms bearing the name "Genetic Laboratories, Inc." in all our metropolitan centers. Significantly, that is the name given to one of the commercial sperm banks recently opened in New York City, whose ostensible mission is simply to provide men with a backstop for voluntary vasectomy.

I have no explanation of why there is not among medical scientists an upsurge of protest against turning the profession of medical care into a technological function; why there is not, precisely today, a strong renewal of the view that the proper objective of medicine is to serve and care for man as a natural object, to help in all our natural "courses of action," to tend the garden of our creation. (Of the two oldest helping professions, the priesthood has long since abandoned magic to bring about extraordinary interruptions of natural processes; today, the mantle of such priestly incantations seems to have fallen on the core of medical researchers.)

A Final Irony

Still there seems to be an evident, simple explanation of why people generally in all the advanced industrial countries of the world are apt to raise no serious objection, or apt at least to yield, to what genetic engineering and the manipulation of ova and sperm, and of embryos, will surely do to ourselves and our progeny. It is a final irony to realize that invasions will now be done on man that we are slowly learning not to do on other natural objects; that natural human "courses of action" will be disassembled in an age in which we have learned to deplore strip mining; that in actual practice minerals and vegetables may be more respected than human parenthood.

The reason will be that the agents of these vast changes are authority figures in white coats promising the benefits of applied knowledge; but the deeper reason is that the agents of these vast changes, defendable step by defendable step, are deemed by the public to be not researchers mainly but members of the healing profession, those who care for us, who tend the human condition. Before it is realized that the objective has ceased to be respect for the unborn patient, it will be too late—and Huxley will have been proved true.

Joshua Lederberg, speaking of how public policy may be determined in regard to clonal reproduction ventured his opinion that this would depend on "the accident of the first advertised examples . . . the batting average, or public esteem of the clonant; the handsomeness of the parahuman product" (The American Naturalist, 1966).

Perhaps one can express the paradoxical and macabre "hope" that the first example of the production of a child by the genetic engineering of its parents' gametes will prove to be a bad result—and that it will be well advertised, not hidden from view. I do not actually believe that the good to come from public revulsion in such an event would retroactively justify the impairment of that child. But then, for the same reasons, neither do I believe that germinal manipulation is a procedure that can possibly be morally justified, even if the result happens to be a Mahalia Jackson.

25
Xeroxing
Human Beings

PAUL A. FREUND

When someone says that we must improve the genetic inheritance of future generations, what exactly is meant by "we"? Who decides which questions, by what standards, subject to what checks? Lawyers are accustomed to differentiating questions of fact and questions of law, to define the roles of the jury and the judge. Here it is necessary to be clear about the difference between questions of science and technology and those of ethics and policy.

The vision of a greater human race, the vision of positive eugenics, might take as its text, ironically, some lines of Walt Whitman: "The pride of America leaves the wealth and finesse of the cities and all returns of commerce and agriculture and show of exterior victory to enjoy the breed of full-sized man, or one full-sized man unconquerable and simple." "Full-sized men"—how do we recognize them (again the unanalyzed "we"), and "simple" men—how do we know their value for survival? It is much easier when we are breeding farm animals, for milk or meat or pulling power. During World War II there was a shortage of penicillin, a newly discovered drug, in the North African theater of operations, and the question arose whether priority should be given to soldiers suffering from battle wounds or to those afflicted with venereal disease. The commanding officer, against the advice of his consultants, decided to give

Reprinted by permission of Science and Public Affairs, the *Bulletin of the Atomic Scientists*. Copyright © 1972 by the Educational Foundation for Nuclear Science.

priority to the venereal disease sufferers, and justified his decision in a persuasive way. The overriding goal of his command was to put the maximum number of troops on the front line in the shortest possible time. The battle-wounded would require a longer time to recuperate, and they posed no danger of infection to others; therefore their claims to the antibiotic must be subordinated. For better or worse, the goals of human existence are rarely so circumscribed and the choices so readily rationalized. And the further we peer into the future—from decades to centuries to millennia—the more problematic become our priorities, the more humble our pretensions to wisdom. Who shall choose, even for the sake of the survival of the species, between artists and scientists, poets and engineers, men of cognition and men of feeling?

Perhaps positive eugenics should be looked at in a closer perspective of time. The most nearly available technique for the controlled engineering of offspring appears to be offered by the prospect of cloning, the asexual replication of a male or female progenitor with no interfusion of chromosomes from a second parent. Since the donor's chromosomes are embodied in all his or her somatic cells, there is a practically limitless supply, subject to the availability of egg cells into which a donor cell can be nucleated for sustenance, either in vivo or (conceivably) in vitro. Thus we may be facing a new kind of nuclear explosion, this one of a biological nature. Whether the progeny will in fact turn out to be replicas of the parent is a question for science; possibly mutations will occur, possibly slight changes in nourishment and environment will make a substantial difference; and particularly in the case of a genius as the forebear, the line between creative genius and hapless incompetence may prove to be a tenuous one. Whether the aggregate gene pool will be impoverished seriously is likewise a basic scientific question.

But the ethical issues go far beyond the scientific ones. What the xeroxing of human beings would do to the fundamental premises of personality, moral responsibility and freedom of will must give us great pause. The mystery of individual personality, resting on the chance combination of ancestral traits, is at the basis of our sense of mutual compassion and at the same time, of accountability. Within the individual the mystery is a wellspring of striving and aspiration. To become a new edition, indeed only a new imprint of a parent, could undermine these expectations and aspirations and produce a breed of passive creatures waiting for the familiar ancestral scenario to unfold, a breed for whom praise, blame, wonder, and fulfillment would have lost their meaning.

Perhaps this is the destiny of the race. Perhaps moral premises will themselves have to alter drastically to accommodate the new technologies. If so, we can at least be clear about it and speculate on what the new morality will be like. Some changes have, of course, already occurred in response to scientific advance. Darwinism presented a crisis for literal religionists, as Copernicus had for medieval theology. But these shifts left it possible to retain the essential moral foundations of human responses to human beings; agnostic humanism is at one with revealed religion in its stress on humility and mystery, and the intrinsic human worth that emerges from these avowals of self-limitation. Indeed, as knowledge has grown the horizons of the still unknown have kept receding.

The issue is not whether the search for knowledge should be curtailed; that way darkness lies. The issue is rather whether individual human beings should be constructed with a set of pre-ordained traits, and indeed whether an indefinite number of such identical products should be engineered. This is an issue that transcends scientific freedom, the freedom to inquire and to know, since it can determine for future generations the very capacity and the will to know, no less than the possession of other traits of thought and feeling that we regard as the essence of the human. At whatever council table such an issue is decided, there should be spokesmen for the future generations in whose behalf we would be purporting to act, spokesmen like the guardian appointed by a court to represent unborn heirs of infant claimants.

Negative Eugenics

More modest proposals, for negative eugenics, do not escape these problems of standards and procedures in decision-making. A good starting point for discussion is the case of the mongoloid baby which was the subject of a symposium at the Kennedy Foundation in Washington in November 1971. At a leading eastern hospital an infant was born and diagnosed as suffering from Downs syndrome (mongolism), together with an intestinal obstruction which would be fatal in a matter of days unless corrected by a relatively safe surgical operation. The facts were laid before the parents, who decided, in the interest of their two normal children, that the baby should not have the operation and should be allowed to expire in the hospital. The pediatricians took the position that this decision was binding on them, since as they understood the law it would be illegal to operate on a child without the parents' consent. And so the infant was permitted to linger, unnourished, until death came within some two weeks' time.

The first point of interest in this case is the nature of the decision to be made. For the doctors it was a legal decision, controlled by the parents' wishes. For the parents it was a decision conditioned largely on the assumption that the care of the child would devolve upon the family. Neither decision can be said to have been made with full understanding. It is true that parents must consent to a form of surgery for a child where there is a genuine choice of treatment; but when the choice is between death and a life-saving intervention, the interests of the child cannot be finally willed away by the parents. There are a number of cases where a hospital, seeking instructions from a court in such a situation, has been ordered to perform an operation over parental objection based, for example, on religious scruples against blood transfusion. Those cases, to be sure, did not present the issue of a mongoloid child; but the doctors' position here professed to be based on a general principle of law. Likewise the parents' decision seems to have been based on the assumption of home care, without adequate exploration of the alternative of institutional care (and its costs). In short, there was lacking any forum where all the interests and possibilities could be explored. When it was suggested in the symposium that a hearing in court should have been arranged, with a guardian appointed to represent the child, objection was voiced that lawyers and judges are not experts in Downs syndrome. That, of course, misses the point, namely that the experts should have their full say and that a disinterested arbiter, taking account of the experts' testimony and also of moral standards of judgment, should make the ultimate decision.

Moral Standards

What are these moral standards? They are not purely private; they can be educed by reasoning from analogy. Is it legitimate to put another human being to death in order to make life more confortable, psychologically and economically, for the survivors? Is the case like that of the care of a terminally ill patient, when extraordinary supportive measures may not be taken and the patient is allowed to take leave of life without hindrance? Are the prospects to the newly born and the aged dying patient comparable from the standpoint of the potentiality of savoring some, if not all, of the wondrous experiences of life as a human being? Is there a point, nevertheless, at which this prospect in the newly born is so attenuated in its range that the offspring should not be regarded as a person (what is medieval theology was termed a monster, not "ensouled")? If the condition could have been discovered prenatally, and if abortion would then have been justified, is infanticide similarly justified? If not, is there an obligation on the part of the state to assume the care of the child when the parents seek to renounce it? Is the case for such a social obligation strenthened where the use and the non-use of prenatal diagnosis tend to correspond to the social class of the mother?

The question about social class is a reminder that lurking throughout the subject is the problem of economic and social inequality and how this factor would relate to the various measures proposed. In legal terms, the problem is the equal protection of the laws—the principle that classification must have a reasonable basis, and that where fundamental human values are involved the classification must rest on a compelling public need.

The problem of equal protection has in fact arisen in connection with negative eugenics. In 1927 the Supreme Court considered a case under a Virginia law that required the sterilization of persons in state institutions who were afflicted with a hereditary form of feeble-mindedness. In a summary opinion, Justice Holmes sustained the law, closing his opinion with the quip "Three generations of imbeciles are enough." It used to be fashionable to quote this remark and add: "Mr. Justice Butler dissented." Today we are more likely to be embarrassed by Holmes' jauntiness. Whether such sterilization will have an appreciable effect on the prevalence of feeble-mindedness, in view of the number of apparently normal carriers, is a question for the biostatisticians. But conceding an appreciable effect, the issue of equal protection remains for moralists and lawyers. Is it acceptable to enforce sterilization only upon those in state-supported institutions—manifestly a limitation reflecting differences in social class. In 1970 the Supreme Court came close to reconsidering the 1927 decision, when it granted review of a Nebraska case involving a law that required the sterilization of mentally deficient persons as a condition of their release from a state institution. Fortunately or unfortunately the law was repealed before the case could be heard by the Supreme Court. It is a safe assumption that the Court would have had more trouble with the problem than did Justice Holmes and his colleagues (save for Justice Butler).

Political Check

The principle of equal protection embodies an important political check, apart from its mandate of distributive justice for those immediately affected. Politically, a control device is likely to receive less thorough consideration to the degree that the class of those affected is small and relatively powerless. A law that bears equally on rich and poor, scions of wealth and wards of the state, is more likely to receive the kind of scrutiny in its enactment and its administration that the gravity of the subject requires.

Similar considerations apply when we consider straightforward proposals to encourage population control. It is attractive to search for the middle ground of incentives and thereby avoid both the weakness of mere persuasion and the heavy force of legal coercion. In a society that reflects a market economy, incentives tend naturally to be economic ones —subsides for family limitation, tax burdens for family expansiveness. Unless, again, the result is to be fashioned along lines of economic class, great care will have to be taken to see that the incentives bear equitably upon all classes.

And so, in the end, whether we deal with positive eugenics or negative eugenics or the limitation of population growth, what is possible to accomplish is more than matched in difficulty by questions of what it is right to accomplish and what are the right means to employ. As Einstein said, physics is so much easier than politics.

26
Genetic Engineering

JOHN V. TUNNEY AND MELDON E. LEVINE

The cry has been raised by many that the impact of science has been too fruitful. It has been raised by some with regard to the nuclear sciences. It might well be reiterated in the near future with regard to the biomedical sciences.

The biomedical sciences have provided man with the increasing ability to modify and to alter human genes. These developments touch upon the most fundamental issues of human life. They portend the ability to reshape man.

Accepted attitudes about the inviolable nature of man's genetic endowment now stand challenged by science. The political impact of this challenge might be just as powerful as the scientific impact. Unless the potential provided by the biomedical sciences is properly understood, the inevitable social response might be fashioned out of fear. When complex problems appear too terrifying or mysterious, some people might seek simple solutions that prove inadequate or improper.

If people become sufficiently frightened—if they feel the need to be rescued from a menace they do not understand—they are more likely to delegate freedoms and less likely to respond with reason. If the polity responds to the scientific community through fear and mistrust, we could witness the erosion of our most precious freedoms. Political alteration—like genetic alteration—might be irreversible. If our personal liberty is ever lost, it might never be recovered.

Consequently, the most important and enduring of our freedoms are linked with the manner in which the biomedical sciences are understood and applied. The issues raised by the biomedical sciences must be exposed to public scrutiny. They must be discussed candidly, openly, and at once.

In approaching these issues, we the authors must of necessity wear several "hats." The first "hat," if you will, is a multifaceted one, fashioned around our own personal backgrounds—our social, philosophical, and ethical beliefs. The second "hat" is a legal one, obtained after studying the law and participating in it over a period of years. The third is a legislative one, obtained from the unique perspective to which we are exposed in helping to propose, evaluate, and create the laws of this land.

It is important to recognize that all political figures wear "hats" of this nature. They are all different, depending upon the individual background and experience of the person, but they influence him as he evaluates and determines policy, especially in an area as sensitive and as potentially explosive as genetic engineering.

Before we discuss the ethical, legal, or legislative view, however, it is important to set forth the most salient aspects of genetic engineering and to indicate our assessment of the state of the art in each of these aspects.

Abortion and Amniocentesis

The technique of amniocentesis—prenatal sampling of the amniotic fluid surrounding the fetus—is frequently used to provide advice on therapeutic abortions. The procedure is relatively safe, but we do not yet have the ability, with amniocentesis, to detect all genetic defects. Within five years most monogenic defects that we understand will be detectable thereby, but even then questions will remain unanswered as to whether amniocentesis affects the eventual intelligence of the child.

Mass Genetic Screening

In this, too, we appear to be on the threshold. The technique is available for many diseases, although not for some others. However, it has already become evident that many people will oppose mass genetic screening— whether of children or adults—for a variety of personal reasons. Some feel it is an invasion of personal rights; others do not want children genetically defective to find out that they are so afflicted. Again, however, the technology is increasingly available.

Monogenic Gene Therapy

Modification of certain cells in terms of their genes, or monogenic gene therapy, we have been advised, has not yet been performed successfully. It should, however, be a possibility for certain diseases within five years. As we gain more knowledge about monogenic defects, the possibility of monogenic gene therapy will become more a reality in broader areas.

In Vitro Fertilization

Both in vitro fertilization and reimplantation in the uterus have been performed successfully in experimental animals. If the research in this area is not seriously inhibited by external controls, the technology for in vitro fertilization and for reimplantation in human beings should be available within five to ten years, or perhaps even earlier. Recently a Yugoslav scientist advised a conference in Tokyo that had developed an instrument for oöcyte (egg) transplant into the uterus.

Cloning

Cloning of frogs, where a replica of an individual is developed from one of its somatic cells, has already been successful. The technology for the cloning of mammals will be available within five years, and, unless research is stopped, the technology for the cloning of human beings might be available within anything from ten to twenty-five years.

Polygenic Gene Therapy

We are very far away from achieving polygenic gene therapy—perhaps 50 to 100 years. Our understanding of polygenic gene defects still is extremely primitive. For a variety of reasons it is considerably more complicated to isolate and trace a polygenic trait than to isolate and trace a monogenic trait.

One of the most powerful arguments presented in favor of employing one or more of the technologies of genetic engineering in the direction of genetic intervention is that man's genetic load is increasing. In other words, the total number of genetic defects carried by man has been increasing. This has been occurring as a result of the increased mutation rate that accompanies population growth and the decreased natural selection rate occasioned by modern medicine and technology. About one child in twenty, for example, is now said to be born with a discernible genetic defect.

The question then arises: Should the human species attempt to employ these new technologies to deal with this increased genetic load? Paul Ramsey of Yale has stated that "it is no answer to say that changes are already taking place in humankind or that men are constantly modifying themselves by changes now consciously or unconsciously introduced. . . ." He argues, as a Protestant theologian, that scientific intervention in this area is a questionable human aspiration, as he puts it, "to Godhood." Regardless of the merits of Ramsey's position, the very vigor with which he defends it suggests the extent to which ethical issues are at stake.

The ethical questions raised by the possibilities implicit in genetic engineering are no less fundamental than the issues of free choice, the quality of life, the community of man, and the future of man himself. Thus, it becomes evident that one's own sense of ethics, one's personal view of right and wrong, one's own standard of conduct or moral code, are essential components of decision making in this extremely sensitive area.

Many political scientists like to believe that political decision making can be objectified, that a process can be delineated by which political decisions are made. Through such a process, it is assumed, decisions and actions can be predicted. The wisdom of these political scientists is questionable for a variety of reasons. One of the most important is the significant subjective component of political decision making—the large realm left to one's own values and ethics.

This realm affects all aspects of lawmaking. It is especially important in any political or even legal approach to genetic engineering. One's own values or ethics must inevitably be brought to bear upon a variety of important questions in this area, questions that can be evaluated only by subjective criteria. In an effort to rationalize some of the issues involved, we will attempt to draw some distinctions and to articulate some criteria for analysis. Some of these criteria have been suggested by others; some are our own. We suggest them not as a definitive list but as a reminder that it will be very important to apply criteria such as these to any legislative or legal analysis of the implications of genetic engineering.

Let us posit a list of ten general considerations suggesting possible ethical distinctions:

First, if we are to engage in any eugenics, negative or positive, we must confront three vital questions that pervade this entire subject: What traits are to be considered desirable? Who is to make that determination? When in the course of human development will the choice be made? These questions cannot be underestimated in their importance to the future of man, particularly when we are considering biological alternatives that might not be reversible.

Second, we must ask whether the genetic engineering or "improvement" of man would affect the degree of diversity among men. Does it presume a concept of "optimum" man? Is diversity important as a goal in itself? Does—or should—man seek an "optimum," or does he seek a "unique"? What would the quest for an "optimum" do for our sense of tolerance of the imperfect? Is "tolerance" a value to be cherished?

Third, we should consider whether it might be appropriate to delineate different biological times or moments—at least in humans—during which experimentation might occur. Do different ethical considerations apply if we attempt to distinguish between experimentation on an unfertilized sperm or egg, a fertilized sperm or egg, a fetus, an infant, a child, or an adult? Might the factors to be balanced in making a decision as to whether experimentation is proper vary at different stages of human development?

Fourth, is there·a workable difference between, on the one hand, genetic "therapy" to correct genetic factors known to cause somatic disease and, on the other hand, genetic "engineering," defined as techniques to alter man in terms of some parameters other than somatic disease? Might it be appropriate to attempt such a distinction in definitions in this emotionally charged area? Might the term "genetic therapy" evoke less emotionally charged reactions than the term "genetic engineering"? Might it, in fact, be preferable to respond more receptively to those areas of genetic work that are primarily "therapeutic"? Or is such a distinction unworkable?

Fifth, it would seem to be advisable to ask whether a particular technique or technology is devised for the therapeutic treatment of an individual or whether it is designed to have a broader societal impact. This potential distinction has a variety of ramifications. For example, it should be asked whether techniques developed for the therapy of an individual patient automatically diffuse into the general public for purposes other than this therapy. Are physicians operationally capable of restricting the use to one group, or does societal pressure make them semiautomatic dispensers of seemingly desirable technologies?

Sixth, we might ask whether any eugenics program—whether positive or negative, voluntary or compulsory—does not imply a certain attitude toward "normalcy," toward a proper norm for human activity and behavior, and toward expectations with regard to the behavior of future generations of human beings. Implicit in this question are distinctions with regard to positive versus negative eugenics programs and also with regard to compulsory versus voluntary eugenics programs.

Seventh, how are words such as "normal," "abnormal," "health," "disease," and "improvement" defined? Are they words that can be operationally used to determine what should be done in the area of genetic engineering?

Eighth, we must ask if the quest for genetic improvement would be continuous. Would it invariably make all children "superior" to their parents? What would be the social consequences of this? Would it institutionalize generation gaps and isolate communities by generations?

Ninth, we should consider whether the institutionalization of a quest for genetic improvement of man is likely to lead to his perception of himself as lacking any worth in the state in which he is. What does this do to the concept of the dignity of the human being in his or her own right, regardless of some "index of performance"?

Tenth, if we have a well-developed ability to perform genetic therapy as an assault upon certain diseases but such therapy is not available for all who have the affliction or who desire the "cure," the question will immediately arise as to how to determine which patients will receive it. Are some classes or groups of people more desirable patients or more worthy of treatment? How will selection be made? By what criteria will those decisions be reached?

Questions such as these ten can be answered only by appealing to ethical, or so-called moral, arguments. When we enter this realm, it is important to remember that no one has a greater claim to wisdom than anyone else. All men have a stake in this area, and all men have a right to be heard.

We would like to offer three additional thoughts that might affect all of the ethical judgements involved. Two are caveats, and one might be a preliminary guide for analysis.

The two caveats are reminders of the imprecision of measurement and the difficulty of meaningful analysis in this area. As for the imprecision of measurement (caveat number one), Ramsey states that "many or most of the proposals we are examining are exercises in 'what to do when you don't know the names of the variables.' " While that might be somewhat harsh, he is accurate in his suggestion that prediction of behavior or even of most genetic disease will be very difficult, owing not only to polygenic factors but also to such other imponderables as pinpointing a recessive trait in its heterozygous state and predicting the influence of environmental factors.

The second caveat is the difficulty of meaningful analysis. Some values we will be asked to compare will be like comparing apples and oranges. How can one, for example, compare the possible deep satisfaction experienced by an infertile woman carrying and bearing a child that was fertilized in vitro and reimplanted in her uterus with the 1 or 2 or 5 per cent chance that the child will be deformed? In measuring eugenic traits to be cherished, how can one compare intelligence (even assuming it can be defined) with love?

In making genetic choices—and in selecting those who will make them —one should not forget such caveats.

The last of these attempts at ethical classification is an effort to ask the question of just where in the broad field of genetic engineering the ethical issues will arise. At what level in the process? Professor Abram Chayes of Harvard Law School has suggested that at least three levels can be discerned at which the questions posed above might arise:

First, the general level of research. Should research be pursued that might lead to technologies that will give science the genetic capability to engineer human beings?

Second, the level of treating human disease. Questions will, of course, be raised as to what exactly is a disease—and how it is defined. (Should socially undesirable or disruptive behavior be treated as an illness? Are some forms of mental illness proper candidates for genetic therapy?) Even assuming that those questions can be answered, ethical considerations will arise with regard to whether the disease should be treated with a newly available genetic technique.

Third, the broad level of attempting to affect society—the level of what some will consider an improvement of the human species.

These levels overlap to some degree with the questions we have already raised, and it appears clear that the ethical pressures that will be directed against the continuation of the activity will become increasingly strong as we move from the first to the second and then to the third level. While these considerations by no means exhaust the ethical realm, they do suggest the enormity of the problems with which we are attempting to deal. Perhaps the attempt—however primitive—at ethical classification might also offer the lawyer some general guidance.

It might be asked, for example, whether there are legal as well as ethical distinctions between negative and positive eugenics. Are there legal differences between an attempt, on the one hand, to treat an individual for disease by either monogenic or polygenic gene therapy and an attempt, on the other hand, to control behavior or otherwise alter society's norms? Does it matter—legally—at what point in the state of human development the therapy or the engineering is conducted: whether in the stage of birth control, in the realm of abortion, in treating a minor or an adult? Does it matter—legally— how the therapy or the engineering is conducted, whether it is voluntary or compulsory, whether for punitive or eugenic reasons, or whether the physician has freely and openly obtained the consent of the patient?

Clearly, these distinctions ought to be important—legally as well as ethically. They touch upon fundamental and traditional legal principles, principles that have been applied in Anglo-American jurisprudence for a number of years. They offer the lawyer a variety of factors that will help in his analysis.

Law, at least in the United States, can be said to operate on three broad tiers or levels. First, we have constitutional law, or the legal framework set forth by the Constitution of the United States and by the courts in interpreting the Constitution. Second, we have statutory law, or law that is enacted by statute—either of a state or of the federal government. No statute can contravene a constitutional requirement. But, in the absence of a constitutional prohibition or pre-emption, federal and state legislatures can enact statutory standards to respond to a variety of needs, such as those that arise in the areas of health and welfare. Third, in the absence of a controlling constitutional or statutory provision, the courts rely upon the body of law known as the common law— those legal principles that have emerged from judicial decisions. The issues raised by the technology of genetic engineering affect constitutional, statutory, and common-law principles. We shall briefly consider each of these legal tiers.

At least three constitutional factors clearly emerge when one considers the general subject of genetic engineering. The first is the right to privacy. The Fourth Amendment to the Constitution declares that "the right of the people to be secure in their persons, houses, papers, and effects, against unreasonable searches and seizures, shall not be violated. . . ." This language has been interpreted to guarantee to the individual a constitutional right of privacy. Genetic engineering raises questions with regard to the extent and inviolability of that right. The second factor involves the rights protected in the Fifth and Fourteenth amendments, which guarantee that no person shall be deprived of life, liberty, or property without due process of law. Third, and perhaps the most important factor that the Constitution brings to bear upon genetic engineering, is the *approach* of constitutional law—the method of analysis that courts have developed for dealing with constitutional issues. Apart from the technicalities inherent in whether state action is or is not involved—a threshold question in any constitutional analysis—constitutional law requires the government to show a more compelling governmental need when the abridgment of fundamental freedoms is involved.

Let us take two examples. Contrast, for instance, a government-sponsored compulsory program of negative eugenics, designed to eliminate a certain genetic disease, with a government-sponsored compulsory program of positive eugenics, designed to control behavior. As both programs are compulsory, both could infringe the fundamental freedom of procreation and possibly of marriage. However, compelling state interest could be advanced as a more legitimate argument in eliminating a disease rather than in controlling or altering behavior. The eradication of disease has long been accepted as a vital social objective. We do not offer this dichotomy in an effort to support the negative eugenics program. In fact, we would probably oppose it. But we do think that a constitutional analysis of the two approaches would bring different factors into being and might yield different results in the two cases.

To move from constitutional law to statutory law, it should be noted at the outset that a variety of statutes in numerous American jurisdictions have attempted to impose eugenics controls. Professor William Vukowich of Georgetown has written: "In the early 1900s, many states enacted laws that prohibited marriage by criminals, alcoholics, imbeciles, feebleminded persons, and the insane. Today most states prohibit marriage by persons with venereal disease but only a few states have laws which are similar to those of the early 1900s. Washington and North Dakota, however, still prohibit marriage by women under forty-five and men of any age, unless they marry women over forty-five, if they are an imbecile, insane, a habitual criminal, a common drunkard, feeble-minded or [a] person who has . . . been afflicted with hereditary insanity."

A number of the more recent developments in the field of genetic engineering, however, go entirely unregulated. Sperm banks—which may be used as a reserve of sperm for artificial insemination by third-party donors, for example—are an excellent example of institutions for which pertinent statutes do not exist. Their administration is entirely up to the persons operating them.

That is an instance in which our third legal tier, the tier of the common law, must be our guide. In the absence of constitutional or statutory guidance, we must turn to the common law for our standards. Here again the law is neither silent nor comprehensive. It falls somewhere in between. Assume this set of possibilities: Amniocentesis is an everyday practice, held by most doctors to the free of harmful effects such as infection. A woman who has not been offered amniocentesis gives birth to a Mongoloid child. Is her obstetrician liable for malpractice?

Common-law tort principles of malpractice would probably hold that the doctor would, in fact, be liable. This is so because the common law in determining negligence tends to follow whatever is the accepted medical practice for a particular community. But is this a viable solution? Would it be appropriate to require amniocentesis even if the mother—or the doctor—has strong religious convictions that preclude consideration of an abortion under any circumstances? What about offering amniocentesis under those circumstances? And what about the legal rights of an egg that has been fertilized and grown in a test tube? Does the *father* have any rights? What rights does the *mother* have? Or the doctor? Do common-law tort or property rights apply to this question?

However one evaluates these issues, they must be faced. If one does not wish to face them with the exclusive guidance of the common law, the result will be the consideration of new legislation. To the extent that current law is inadequate, legislation must be developed.

In considering the possibility that legislation must be developed, we both are painfully aware of the potential inadequacy of the legal and the legislative processes in responding to issues presented by science. In the area of genetic engineering, science may be outpacing the legal and legislative processes. It may be presenting challenges to which our lawyers and legislators are ill-equipped to respond. Our legislative system may be poorly equipped to respond to these problems because of at least two inherent difficulties: its speed and its scope.

Our legislative process generally works slowly. Sen. Walter Mondale, for example, first introduced legislation that called for a commission to study the effects of genetic engineering almost five years ago. That bill passed the Senate unanimously last year, but it has not yet been acted upon by the House of Representatives. Just initiating study commission on so momentous a subject has already, then, taken longer than five years.

Not only is our political system slow. It is, obviously, only national in scope. Generally, that is not a significant problem to the people in Washington who are considering various legislative proposals. Most proposals are only national in scope—or less. Genetic engineering, however, is clearly a matter of international concern. It will require, if any controls or guidelines are to be effectively suggested, international agreements. This also will serve as a political or legislative constraint.

Recognizing these constraints, we still believe that certain constructive steps can be taken—steps that will begin to offer legislative rationalization to the field. If the legislative system begins to consider these problems now, it might be possible to respond politically and legislatively before it is too late.

Conversely, we fear the consequences that could be wrought if informed legislative consideration of the issues inherent in genetic engineering does not soon begin. If the legislation comes as a result of dramatic scientific breakthroughs that scare the public, the outcome might be hasty and unwise political decisions predicated upon inadequate information and upon fear. If debates and discussions begin now, however, the ultimate legislation might emanate from deliberate and reasoned political, social, and scientific analysis. We do not believe that, at this point, it would be appropriate to suggest answers to the momentous issues raised by genetic engineering. But we do believe that we know enough to undertake certain legislative initiatives. Let us suggest three.

First, Congress should enact the Mondale bill [S. J. Res. 75], which provides for a study and evaluation of the ethical, social, and legal implications of advances in biomedical research and technology. The proposed study commission might serve as a preliminary vehicle for educating the public about the foreseeable social consequences of biological advances. Such a commission might best be an international one, but that is logically a second step.

Second, and perhaps equally important, is the initiation of technology assessment in all institutions that disburse funds, direct research, or provide grants that are related to biomedical concerns. It has long been obvious that technological developments have implications that affect society in a variety of ways and that their impact cannot be limited to an analysis of the technical aspects of the product or of the innovation. Similarly, the myriad implications of the developments of biomedical research reach out to all segments of society. Technological assessment should be a part of any analysis of any project that involves a potentially new biomedical development.

Third, it might be appropriate for Congress to earmark a small proportion of health research funds (say one-quarter or one-half of 1 per cent) for research into possible social consequences of biological technologies either presently availalle or foreseeable.

It is not only the legislature, however, that can initiate improvement in communication between the scientific community and the general public as well as expansion of public awareness of and concern with these issues. Four other suggestions might be worthy of consideration:

First, private foundations should be urged to initiate programs to bridge presently existing gaps between the sciences and the humanities, exposing people in each area to people in the other, and making the ideas of each readily available and understandable to the other.

Second, universities should consider establishing additional programs whereby students in the humanities would be exposed to the methodologies familiar to those in the scientific disciplines, and vice versa. The two general groups should feel a closer realtion to and understanding of each other in universities as well as elsewhere. The effective separation of these two groups in universities—particularly at the graduate level, but even at the undergraduate level—establishes a line of demarcation between those in the sciences and those in the humanities, with very inadequate and narrow bridges to unite the two general areas.

Third, research proposals in the biological area should perhaps be assessed by institutional research review committees that include non-scientists. Some form of technological assessment, in other words, or consideration of the ethical, moral, and social implications of biological projects—by non-scientists—should be considered at the level of all research proposals in this general area.

Fourth, it might be appropriate for the medical profession itself to study the ways in which the technologies it uses for the benefit of individual patients may affect society as a whole if used for purposes other than the cure of individual patients. The "individual treatment versus social engineering" dichotomy should be considered clearly and carefully by the medical profession and should probably be emphasized more strongly than it currently is.

These efforts to bring society and the biomedical sciences closer together are, in our opinion, essential. Dr. Andre Hellegers, director of the Joseph and Rose Kennedy Institute for the Study of Human Reproduction and Bioethics of Georgetown University, has testified before the Senate that "nothing could be worse than that society should come to fear scientific progress. . . . I can foresee that the occasional, seemingly sensational, scientific episode will so frighten society as to undermine the very support [that] science needs in order to continue to make contributions to improve the lot of mankind. . . . It is high time that there be started an educative process that explains to the country the precise nature and limitations of the scientific process and the place it occupies in man's control of his environment. . . . No segment can stand apart in this interdependent society. If it attempts to do so, it is bound to cease being supported. The sooner the relationship of science to society is examined and explained for all to see, the better it will be both for science and for society."

This testimony touches upon two very important facts of American political life, neither of which should be forgotten. First is the theory of political accountability; if the public supports something financially, the public is entitled to know what it is that it is supporting. Second is the foundation of political democracy; thoughts, suggestions, proposals, and policies should be scrutinized in the market place of ideas. Political debate and public discussion are healthy and are conducive to the best analysis of any position. Particularly in an area as fraught with subjectivity as this one, it is vital that the issues raised be aired, discussed, and debated. We are dealing in an area in which there is no monopoly of expertise. Rather, it is a field in which men trained in a variety of different areas, or even in no special area, bring to bear their own unique perspective, or, if you will, "expertise." We are dealing with a subject in which morality, or one's own subjective sense of ethics, is pervasive. We are, therefore, dealing with an area in which all persons have a right and a special claim to be heard.

There are certain suggestions that we would offer in any debate on this subject. We would suggest that among the values that man ought to protect most fully are the values of humility, of compassion, of diversity, and of skepticism. We would suggest that any scientific or technical initiatives of one generation that would foreclose or eliminate the options of future generations—any decision today that implies an ability to predict the human traits that will be most cherished tomorrow—smacks of arrogance and should be avoided. We would suggest that man should exercise the utmost caution in this sensitive field and that decisions that will be genetically irreversible might require a wisdom we do not possess. We would also suggest that there is no reason why the ethics or morality of any one of us is better than that of any other. In the realm of morality each of us has an equal claim to wisdom.

Therefore, the issues raised by the biomedical scientists must be debated, and the debate must begin now. If we postpone debate in this area, we might face irreversible trends not only in genetics but also in political freedoms.

All segments of society should be involved in the debate these new technologies demand. The techniques must be discussed and debated among lawyers, doctors, theologians, legislators, scientists, journalists, and all other segments of society. The issues raised require interdisciplinary attention. We cannot begin too soon to consider them.

27
Moral and Legal Decisions in Reproductive and Genetic Engineering

WERNER G. HEIM

Man has long known that he can modify the genetic bases of organisms by controlling their breeding patterns. The application of such knowledge to the development of specific breeds of dogs or of wheat strains particularly suitable to cultivation is quite old. Mendelian and post-Mendelian genetics increased this ability by providing a rational model for the observed effects and thereby increased the ability to predict. This ability, when combined with advanced knowledge in the physical sciences, led to an understanding of many of the major mechanisms for the genetic transmission of information. Finally, this newest knowledge is rapidly providing the means for the manipulation of that hereditary information itself.

Imminence of the Developments

The ability to change the hereditary information content of a human cell is not a matter for the future: it is presently available. In 1971 Merril, Geier, and Petricciani showed that a gene—a unit of hereditary information—from a bacterium could not only be introduced into human cells but could be caused to function in them. Nor need we any longer simply search for a gene that we may wish to manipulate: in 1972 a human gene was synthesized by Kacian and his associates. These techniques are still laboratory exercises; but the time-span from first demonstration to

Reprinted, with permission, from *American Biology Teacher*, Vol. 34, No. 6.

practical application is not likely to be longer than for similarly important developments in the physical sciences. Transistors, for instance, were in general use within less than a decade of their initial development.

An additional line of research is likely to accelerate the application of genetic manipulation, or genetic engineering, to man. This is what may be termed reproductive engineering. Reproductive engineering is any deliberate manipulation of the procreative part of the life cycle. Much of this is already in daily use in such forms as conception control and artificial insemination.

Genetic and reproductive engineering tend to be synergistic. Consider, for example, the production of genetically mosaic mice; that is, mice whose heredity is literally a patchwork. This condition is achieved by fusion of very early embryos of different strains. This requires at least four interacting techniques: (i) the exact timing of pregnancies, (ii) the manipulation of the eight-cell embryos in vitro, (iii) the production of exactly timed pseudopregnancies in the recipients, and (iv) the safe implantation of the mosaic, tetraparental embryos into the pseudopregnant females. The technique of transferring very early embryos from the uterus of one female to the uterus of another—this is called inovulation—itself has implications, to be discussed later. Another potential synergism between the two kinds of engineering may soon appear in the form of the insertion of genes into sperm prior to the use of that sperm in artificial insemination.

A warning should be added here, however. In all of these processes highly specific techniques must be used. This means that one kind of manipulation may be possible at a particular time while another, and apparently closely related, procedure remains out of reach. For example, it is now possible to insert a gene into a human cell; but to remove a specific gene is not yet possible, and indications are that it may remain impossible for a long time. It follows that only a person well acquainted with both the scientific literature and the conceptual basis of these activities is in any position to estimate the time scale for a future development.

Uniqueness of the Developments

These new scientific developments, which have been and are being incorporated into the technology of our society, pose serious moral and legal questions. That, of course, is not in itself new; scientific developments, when more or less broadly applied, have always raised such questions. (Recent examples include the fluoridation controversy and the legal aspects of artificial insemination.) What is new is the nature of the questions raised by the development of genetic and reproductive engineering: the manipulations of man, in these fields, will be vastly more fundamental than any previous ones. In some instances the manipulations will be irreversible, not only with respect to the individual but also with respect to all his descendants and to the population. In such circumstances wrong decisions can lead not only to the death of the individual but also to extinction of the species. Other incorrect decisions could seriously degrade the quality of life through control of thought, of liberty, and of motivation.

Roles in Controlling the Applications

Unfortunately the track record of those who, in my opinion, should be developing the moral positions, public policies, and legal constraints concerning these developments is not good. With few exceptions the moralists, theologians, sociologists, political scientists, legislators, legal scholars, and judges, whose job it is to integrate new developments into a sustained, viable, and sound fabric of society, have not only failed to be prepared for the introduction of new, science-derived technologies into the culture but have even reacted, at best, rather sluggishly. Their reaction time has sometimes been slower than that of the general public, as is shown by the widespread acceptance of certain developments long before they are placed in a definite legal and moral framework. An example is the debate about "when and whether to pull the plug" of the machine keeping a terminal patient more or less alive. This debate commenced seriously only well *after* the machines had come into common use and is still often being conducted in terms hardly useful to the person who must make the decision: the physician on the ward. Another example of the failure to develop an adequate set of moral stances has to do with genetic counseling and the individual and societal risks arising from the reproduction of persons who are sustained by medical skills in the face of hereditary defects. Should an 11th commandment have been pro-mulgated: "Breed not, ye who carry defects"?

One popular reaction to this lack of guidance has been a feeling that research in these fields should be stopped because "it's too dangerous." Of course it is not the research itself that is dangerous but the application of the results within a society that has developed neither the broad principles nor the specific directives to make these applications wisely.

By default, the natural scientist himself has had to fill the vacuum with his own inexpert opinion so frequently that the impression is now abroad in some circles that this, too, is a part of his job. The job of the natural scientist is to make the discoveries; that of the technologist is to develop applications; and that of the social scientist and humanist is to suggest whether, how, and under what conditions the work of the other two ought to be applied.

As a well-informed citizen the natural scientist should have, of course, the same opportunity and at least the same responsibility to contribute to the decision-making process as any other well-informed citizen. But he is not an expert in, and should not be expected to act as if he were an expert in, the delicate processes of weighing the data or of drawing conclusions and making recommendations. He does, however, have responsibilities in this constellation of events beyond those of the ordinary person. First he should give plenty of advance notice to the humanists and social scientists of forthcoming developments likely to require their attention. In doing so he must put all his expertise to work to differentiate between science and science fiction, lest the world react like those of whom the shepherd cried "wolf" too often. Second, he must be prepared to give the relevant details of new developments to his colleagues in the humanities and the social sciences. He must do this with due regard for (i) relevancy, lest he either swamp or impoverish the communication channels; (ii) scrupulous accuracy, lest he mislead his listeners; and (iii) intelligibility of language, lest he be misunderstood. Finally, he ought to monitor the tentative and the definitive pronouncements of the humanists and social scientists so as to detect early any misinterpretation or lack of facts that may have distorted their work.

The need for careful formulation of moral and legal positions on new developments *before* their widespread use is now more critical than ever before: the changes are more fundamental in nature, are less likely to be reversible in the individual or in his descendants, and, most importantly, are changes in human nature itself. As Leon Kass (1971) has pointed out," . . . both those who welcome and those who fear the advent of 'human engineering' ground their hopes and fears in the same prospect: *that man can for the first time recreate himself.* Engineering the engineer seems to differ in kind from engineering his engine." (Kass's emphasis.)

Aspects of Human Engineering

The "human engineering" of which Kass speaks—it is inclusive of what is here called genetic and reproductive engineering—differs from previous kinds of changes in several respects. Consider its influence on human conduct. The traditional methods of modification have had three important characteristics: (i) they used symbols, especially as embodied in speech and art, as their primary vehicle; (ii) they allowed considerable choice to the individual as to the acceptance of at least parts of the modifications offered; and (iii) their effects could, to a great degree, be reversed in both the individual and his progeny. In contrast, the changes brought about by human engineering are largely nonsymbolic, because they modify the conduct-controlling mechanism directly. For example, the human engineer would *not* seek to educate or train a victim of Down's syndrome (the so-called mongolian idiot) to the point of self-sufficiency but rather would eliminate the chromosomal defect or, alternatively, would block fertilization or development of eggs carrying the defect. This is, incidentally, in sharp contrast with one of the most advanced of the old-style techniques, Skinnerian operant conditioning.

The second contrast between the old and the new techniques lies in the fact that the individual frequently will have no power to adopt or reject a particular modification. The modification will have occurred before he became an individual—either in the gametes or the early embryo that produced him or in an ancestral generation. To the extent that the change is in the hereditary material itself, it may be an irreversible change for either of two reasons: the technique for reversing it may not be available—at present a gene may be added but not subtracted—or else the desire to make the reversal at some future date may have been blocked genetically at the same time that the other changes were made.

A further consequence of the difference between the old and the new techniques is that the new techniques have the effect of removing literature, mythology, religious liturgy, etc., one more step from the real seat of power. Once, long ago, a speech—a curse—was expected to kill an enemy. Later a speech was expected to inspire soldiers to kill the enemy. In the future a speech might cause technicians to change the genes of some persons so that, under certain conditions or at a certain age, these persons simply die.

"Now" and "Soon" Examples

What, then, are some of the more likely of these human-engineering techniques? I wish here to deal only with those that are either available right now or have a great likelihood of becoming available in the next few years. Due to the highly technical nature of the work, long-range prophecy is likely to be unprofitable.

One technique that fits the human-engineering category, although it is neither reproductive nor genetic, is that of the direct control of man's neural system by electrical or drug stimulation of specific parts of the brain, as detailed so well by Delgado (1969). One need only read his book to discover the tremendous power available here and the need to develop a proper framework for that power.

Other nonreproductive and nongenetic methods are those devised for the prolongation of life by what, at any point in time, would be considered "extraordinary" means: heart transplantation, heart prosthesis, and the like.

In the realm of genetic engineering itself, one must at least mention genetic counseling—already widespread and rapidly spreading—because serious moral problems crop up almost daily in this work. If a prospective child has a risk of 5% that he will be seriously defective, should the parents reproduce? Or, if conception has occurred, should an abortion be done? What if the risk is 10% or 30% or 60%? What if the risk is essentially zero that the child will be affected but quite substantial that the child will transmit the deleterious gene?

Let us turn now to questions of genetic engineering in the strictest sense. Here we are dealing with techniques that change "human nature" and are transmissible directly to the offspring. Under what conditions should such techniques be used? The technique for inserting a gene or a group of genes into sperm is likely to be practical within a very few years. Suppose two persons are defective in the gene for the production of the enzyme parahydroxylase: their child may suffer from the disease phenylketonuria (PKU). It should be possible soon to introduce the normal, active gene into the sperm of the father and thereby assure that the cells of the offspring have, each, at least one properly functioning gene for this enzyme. Should this procedure be done? Consider not only that normal offspring will be produced but that both of the defective genes will be available to, although perhaps not active in, further progeny.

In the eugenic uses of genetic engineering that I have mentioned here, most of the arguments seem to be on one side: because it is highly likely that both the individual and society will benefit from the engineered changes, and because either short-term or long-term deleterious effects are unlikely, the judgment probably will be that the procedure should be carried out. But let's take a less simple case. There is good evidence for a hereditary component in the origin of schizophrenia and schizoid conditions (Heston, 1970). Very likely we are dealing with a multigenic inheritance pattern here. Therefore there may be a defective gene that facilitates the appearance of a mild schizoid condition. It is not unreasonable to suppose that it should become possible to insert a gene into the fertilized egg—a gene that would "correct" the defective gene. At first sight this would seem to be another clear-cut case for the use of the technique. But there are reasons to suppose that mild schizoid tendencies are useful in a wide range of endeavors, from painting and composing to scientific research. It is obvious that, without further information, both the long-term and the short-term benefits cannot be weighed properly against the possible harm. And remember that once such a corrective gene has been introduced, it may not be removable! In such an equivocal situation what ought to be the rules for the application of the technique? I submit that, in such a situation, the role of the biologist is to provide the raw information and, perhaps, an indication of the probabilities of the various consequences. Then the humanists and social scientists should formulate the rules, subject to revision as more information or better techniques become available.

The other type of human engineering—reproductive engineering—also will require much thought and study. Although less likely to have permanent or irreversible effects, its points of action are even more closely bound up with our general ideas of morality and proper conduct than are those of genetic engineering. This is evident from the arguments raging around the two early forms of reproductive engineering presently in use: conception control and artificial insemination. Neither legal nor moral opinion has been stabilized or universalized in respect to these. What are and what should be the legal and moral positions vis-à-vis the following procedures, all but one of which have already been carried out in nonhuman organisms?

1. Artificial insemination with sperm from donors selected for particular qualities.

2. Artificial inovulation in which an egg produced by the wife and fertilized by the husband is transferred to a foster uterus.

3. Artificial inovulation by transfer of an egg produced by a donor, and fertilized either by the husband or a sperm donor, into the uterus of the wife.

4. Insertion of the nuclei of ordinary body-cells into eggs whose own nuclei have been removed, producing thereby any desired number of individuals with exactly the same genetic constitution as that of the donor of the nuclei—a procedure called cloning.

5. Cloning by causing ordinary body-cells to act as if they were fertilized eggs, thereby (again) producing any desired number of persons having the same hereditary makeup as the donor of the cells.

So little thinking has been done about the last two of these possibilities (outside the realm of science fiction, anyway) that we can hardly even list the individual or social advantages and disadvantages that might accrue from having a large number of genetically identical individuals. We are even less in a position to give these possibilities the calm and deliberate weighing that ought to lead to acceptable patterns of applicaiton.

Prospects and Challenges

The following general conclusions can be drawn:

1. Sepcific modification of individuals, by action on the genes or on the reproductive patterns and for individual or social purposes, is now possible and is increasingly and rapidly becoming feasible.

2. These techniques differ fundamentally from older approaches based on restricted breeding patterns—classical eugenics and Hitlerian eugenics, for example—in two critical respects: they are fast and they work. Even the insertion into man's heredity of what might be termed socially specific genes is a rather close—an uncomfortably close—probability.

3. We need to develop ways of adjusting these possibilities, of restricting their use within morally and socially acceptable patterns.

4. To do this will require a rejuvenation of the humanities and the social sciences and a reshaping of the relationship between them and the natural sciences.

Athelstan Spilhaus (1972) has said: "Just as technological invention cannot remove the need for social invention, neither should our slowness in changing outmoded social practices, institutions, and traditions be allowed to slow technological realization of potential benefits to all." Unless we rearrange our houses so that we can get to work immediately and unless we do get to work immediately, the genie (or gene) will be out of the bottle before we know the magic formulas that control it. And to permit this kind of genie to be out of control or to be misused is (to use the words in an unusual but meaningful combination) to *commit extinction*—first of freedom, and then of the species.

As teachers we have additional duties: to inform our students of these developments and to help them make the questions raised by these developments a prime order of business in their lives. We may confidently expect their lives to encompass the period during which most or all of the developments outlined here, as well as others, will be brought into practice. Those who are presently our students will be the ministers, the judges, the political scientists, the philosophers, the legislators, the teachers and, yes, the biologists, in charge. We and they had better start to think fast about how to handle that charge.

References

Delgado, J. M. R. 1969. *Physical control of the mind: toward a psychocivilized society.* Harper & Row, New York.

Heston, L. L. 1970. The genetics of schizophrenic and schizoid disease. *Science* 167: 249-256.

Kacian, D. L., *et al.* 1972. In vitro synthesis of DNA components of human genes for globins. *Nature New Biology* 235: 167-169.

Kass, L. R. 1971. The new biology: what price relieving man's estate? *Science* 174: 779-787.

Merril, C. R., M. R. Geier, and J. C. Petricciani. 1971. Bacterial virus gene expression in human cells. *Nature* 233: 398-400.

Mintz, B. 1971. Genetic mosaicism in vivo: development and disease in allophenic mice. *Federation Proceedings, Federation of American Societies for Experimental Biology* 30: 935-943.

Spilhaus, A. 1972. Ecolibrium. *Science* 175: 711-715.

28
Genetic Manipulation: Temporary Embargo Proposed on Research

NICHOLAS WADE

Because of a remote but possible hazard to society, a group of molecular biologists sponsored by the National Academy of Sciences (NAS) has called for a temporary ban on certain kinds of experiments that involve the genetic manipulation of living cells and viruses. This is believed to be the first time, at least in the recent history of biology, that scientists have been willing to accept any restrictions on their freedom to research, other than those to do with human experimentation.

The group's proposals, published in this week's editions of *Science* (page 303) and *Nature,* take the form of an appeal to colleagues throughout the world that they follow the group's members in volutarily deferring for the time being two related types of experiment, and in exercising caution before proceeding with a third.

In addition, the group, known as the Committee on Recombinant DNA Molecules, suggests that the National Institutes of Health (NIH) appoint a committee to give practical guidance on the situation, and that an international meeting of scientists be convened early next year to discuss how to proceed further.

The group is chaired by Paul Berg, chairman of the Stanford University department of biochemistry, and most of its members are scientists who are already active in, or have considered entering, the research field in question.

Science, 1974, Vol. 185, Issue 4148, pp. 332-334. Copyright 1974 by the American Association for the Advancement of Science.

The primary object of the proposals announced this week is to buy time for further thought before the rapidly developing research field grows too large to be controlled. It is unclear how far and for how long the appeal will be heeded.

Also uncertain is whether the ban will be observed by countries interested in the new technique's considerable potential for biological warfare. Many millions of dollars were invested at the U.S. Army's biological warfare laboratories at Fort Detrick, Maryland, in try-ing—without much success—to improve on the lethality of viruses and bacteria harmful to man. The new technique offers a theoretically possible way of accomplishing precisely that.

The motivation of the Berg group's proposals springs not from any long-range misgivings about biological warfare or the social impact of genetic engineering, but rather from direct concern about the health hazard presented by the genetically altered bacteria that are created with the new technique. The group recognizes that adherence to its recom-mendations "will entail postponement or possibly abandonment of certain types of scientifically worthwhile experiments" but says that its concern "for the possible unfortunate consequences of indiscriminate application of these techniques motivates us to urge all scientists working in this area to join us in agreeing not to initiate experiments of Types I and II . . . until attempts have been made to evaluate the hazards . . . ."

The new technique, briefly, depends on the use of a newly discovered class of enzyme to introduce particular genes of other species into living cells such as bacteria. The two types of experiment the Berg group says should be eschewed are those that involve inserting into bacteria (i) bacterial genes which confer either resistance to antibiotics or ability to form bacterial toxins and (ii) the genes of viruses. The potential danger—and it is no more than a theoretical possibility—is that the bacteria endowed with such genes might escape and infect the population, particularly since the standard bacterium used by molecular biologists is *Escherichia coli,* a common inhabitant of the human gut.

A third type of experiment—inserting animal genes into bacteria—is one that the group says "should not be undertaken lightly." This is the recommendation that may cause the most controversy. Some scientists consider the group should have proposed a ban on this type of experiment as well, yet those working in the field consider it the key to significant and immediate discoveries, such as elucidating the structure and working of animal chromosomes.

The Berg group's announcement is apparently the first time that biologists have publicly called attention to the possible public hazards of their own research since November 1969 when a team of Harvard scientists—Jon Beckwith, James Shapiro, and Larry Eron—warned of the dangers of government use of science on the occasion of announcing their isolation of a pure gene from a bacterium.

The Berg group's suggested embargo on gene insertion experiments is also remarkable in that it is the first time—at least within the memory of many people consulted by *Science*—that researchers have ever suggested that their own line of investigation should be halted. Others view the proposed halt less dramatically as just an extension of the existing restrictions on human experimentation. According to Berg, the embargo is "the first I know of in our field. It is also the first time I know of that anyone has had to stop and think about an experiment in terms of its social impact and potential hazard."

Will the Embargo Stick?

How will the group's recommendations be received by the scientific community? The proposals have been discussed with colleagues and at several scientific meetings with apparently favorable response. "I think the recommendations will stick," says group member David Baltimore of MIT, "because they are reasonable, and the better part of the scientific community recognizes the need for caution. The worse part will be under a kind of moral pressure to go along with the majority." According to Berg, "Anybody who goes ahead willy-nilly will be under tremendous pressure to explain his actions."

Others are less optimistic that those not represented on the group will blindly follow its recommendations. "Caltech and Harvard will respect them, but those not in the elite will see no reason to hold off," says the editor of a biological research journal. "Anyone who wants to will go ahead and do it," is the verdict of an NIH scientist, who adds that although the technique requires a moderate degree of sophistication at present, it will be a "high school project within a few years."

A preliminary tasting of opinion suggests that the majority of scientists will firmly endorse the group's recommendations for a temporary ban on type I and II experiments (introduction of other bacterial or viral genes into a bacterium). But at least three kinds of objection can be expected from responsible critics.

First, there are those who believe the ban should have been extended to cover type III experiments (introduction of animal genes into bacteria).

Virologist Wallace P. Rowe, of the National Institute of Allergy and Infectious Diseases, considers that such experiments should only be done when bacteria are found that are quite unable to infect man. Another NIH scientist, Robert G. Martin, feels the Berg group should have recommended "complete abstinence" from type III experiments. (Martin, who is chairman of the NIH biohazards committee, stresses that this is his personal, not official, opinion.)

A second source of criticism may be scientists who believe that type III experiments present no hazard to health. Donald D. Brown, for example, of the Carnegie Institution of Washington, plans to put into bacteria the set of silk moth genes that govern the synthesis of silk protein.

"I cannot see how this could cause any conceivable danger to anybody," says Brown who, however, supports the embargo on type I and II experiments. The Berg group is concerned about the introduction of animal genes into bacteria because some animal cells possess the genetic instructions for tumor viruses. Brown, however, believes that any putative viruses inserted into bacteria as a by-product of inserting the silk protein genes will not be a serious problem. (To the objection of one scientist that someone infected by a silk gene-containing bacterium might end up with a "gut-full of silk," Brown replies that, because of differences between bacterial and animal cells, the bacterium would probably produce only copies of the silk gene itself, not of the protein the gene specifies.)

Other type III experiments already carried out include the introduction into bacteria of frog genes and genes from the geneticists' workhorse, the *Drosophila* fruitfly. The promise of these experiments, particularly those involving work with *Drosophila,* is so great that those involved in the area are likely to resent any suggestion that the group's admonition of "carefully weighing" their plans be construed to mean they should be stopped or even postponed.

A third kind of objection may come from those who fear that the mere formalization of such proposals will lead to further impediments on research and remove the ultimate decision from the hands of scientists. "The underlying purpose is excellent," says Joshua Lederberg of Stanford University, "but there is already such a momentum toward the regulation of research that the proponents should carefully consider the consequences of stating such recommendations." A quite contrary view is that of Beckwith, who says he is "happy to see this precedent set because it will raise a debate about academic freedom to pursue whatever research one wishes."

At first appearance the Berg group is vulnerable to portrayal as the fox set to guard the chicken coop. The group's unfoxlike recommendations are evidence to the contrary. Moreover, the group's statement, although endorsed by the NAS, is intended to be the personal appeal of the signatories, a fact which may mitigate criticism that the group has too narrow a membership.

The history of the Berg group's proposals stems directly from Paul Berg's own dilemma as to whether to proceed with an experiment of the type he has now forsworn. Two years ago he managed to synthesize a hybrid DNA molecule which contained a monkey tumor virus named SV-40. Though designed for other purposes, the hybrid molecule would have produced interesting results if introduced into the human bacterium *E. coli*. The problem then arose: what if the *E. coli* containing the monkey tumor virus should escape and infect the population at large?

"I was the first to whom concerns like these were expressed," Berg said last week. "At first it got my back up, but eventually I decided not to do the experiment because I couldn't persuade myself that there was zero risk."

Berg's technique for creating his hybrid molecule was fairly sophisticated but scientists at Stanford and at the University of California, San Franciso, soon developed a simplified adaptation of the technique to introduce frog genes into *E. coli*. "That upped the ante," says Berg, "because it showed how simple it was to introduce any gene you liked into bacteria."

One of Berg's colleagues at Stanford, Stanley N. Cohen, had earlier used the simplified technique to join together two bacterial genes that confer resistance to antibiotics. Cohen described this experiment at a Gordon conference last year. His talk provoked a debate about the implications of hybrid DNA molecules, as a result of which the chairpersons of the session, Maxine Singer of NIH and Dieter Soll of Yale, were instructed to write to the NAS asking that the academy appoint a committee to consider the question. The letter was also published in *Science* (21 September 1973, page 1114).

NAS president Philip Handler directed an academy staff member, Leonard Laster, to follow up on the matter, and Laster asked Singer what the NAS should do. Singer said they should ask Berg ("He was always asking questions about this, and never brushed these questions aside," she told *Science* last week), and Berg replied that he would like to consult others before tendering the academy advice.

With Laster's assert, Berg then got together a group of colleagues who met at MIT this April and agreed that the problems of using the restriction enzyme technique should be put before an international conference, and that the NAS should be so advised.

The conference, however, could not be convened before next February, and meanwhile the new technique was being taken up so rapidly that it seemed many "bad molecules" might have been created before the conference could take action. Berg and his colleagues therefore drafted as a personal appeal the letter that appears on page 303 of this issue. The letter was subsequently accepted by the NAS as representing the committee's report. The NAS endorsement gives the statement an even stronger right to a hearing, though does not in fact mean that the group is claiming to speak for anyone other than themselves.

Prospects for Genetic Engineering

The new technique is a major step toward genetic engineering, since it renders genes accessible and manipulable in a way that has been impossible hitherto. The technique opens up a host of scientifically interesting possibilities. Practical applications remain at the same time very near and very far. One that is often talked of is the posibility of manufacturing insulin by isolating the relevant human gene and adding it to bacteria, which could be cultured and harvested. With use of the restriction enzymes, it looks as if it should be possible to slice up the genetic material of human cells into fragments containing a few genes each, insert the fragments into bacteria, grow thousands of colonies of bacteria, and select the one which contains the gene for insulin (or more precisely, for the protein from which insulin is derived). But though there may seem to be no theoretical obstacles to such a procedure there are numerous practical problems which are nowhere near solution. For a start, it is not known how well, if at all, the genes of higher cells will be transcribed and translated by bacteria.

The utility of restriction enzymes is that they snip the enormously long DNA molecules of living organisms into manageable fragments which are of roughly the order of a gene in length. (This is because the specific sequence of bases at which each enzyme acts tends on the average to occur this distance apart). A second important feature is that some restriction enzymes, when they cut the double-stranded DNA molecule, slice one strand a few bases lower down than the other, leaving what are known as "sticky ends." Since any species of DNA cut by the same enzyme will have the same kinds of sticky ends, the lower part of one

DNA molecule will stick back equally well onto the upper part of another molecule. This is the basic trick whereby two different species of DNA can be annealed into a hybrid molecule.

The way the hybrid is introduced into bacteria is to choose as one of its members—the other is the gene to be inserted—a piece of bacterial DNA known as a plasmid. The plasmid DNA is able to enter the bacterium and get itself (and its hybrid partner) replicated by the bacterium's machinery.

Whatever the prospects for genetic engineering, this is not the reason for the group's suggested embargo. "It is not to be lumped with the proposals saying, 'This is a research path down which we cannot tread because we can't live with the information we will get,'" observes NAS president Handler. The embargo is quite narrowly focused on the specific health hazards potentially raised by genetically altered bacteria, and is framed so as to command the maximum possible agreement among the scientific community. Quite possibly the embargo will be observed until the conference in February. Its real test will come when and if the conference decides the hazard is substantial enough for the embargo to be indefinitely extended. It could then become apparent that control of the new technique is not much easier than the containment of nuclear weapons.

V
POSTSCRIPT:
GENETIC
ENGINEERING
Brave New World?

In the first article in this book you were given the opportunity to examine science writer Caryl River's contention that genetic engineering portends a *grave* new world. You have since read papers by scientists and physicians who have described in some detail significant advances in genetics and how these advances are being or may be employed to correct or ameliorate human genetic defects. In other articles sociologists, legal scholars, and theologians have discussed the social, ethical, and legal implications of such use of recent genetic discoveries.

It is appropriate then, that this book be concluded by returning to the basic question, "Do advances in the study of genetics and reproduction and their application to humans portend a grave new world?" This is the subject of Dr. Arno G. Motulsky's extensive paper "Brave New World?," which appeared in *Science* in late August, 1974.

Dr. Motulsky, director of the Center for Inherited Diseases and professor in the departments of medicine and genetics at the University of Washington, Seattle, provides a summary review not only of different categories of genetic advances, but also of the ethical, social, and legal problems inherent in the application of each to humans. For example, he discusses genetic counseling, intrauterine diagnosis, genetic screening, cloning, and gene therapy. For each of these (as well as for others) he summarizes the current status of the art (science?) and the unique bioethical problems each poses. Finally, Professor Motulsky answers his own title question with these concluding words:

> The ethical human brain is the highest accomplishment of biologic evolution. By harmonizing our scientific, cultural, and ethical capabilities, the potentially achievable results can place us at the threshold of a new era of better health and less human suffering.

29
Brave New World?

ARNO G. MOTULSKY

The public media in the last few years have been full of articles about research in molecular biology and genetics. These fields have come to interest the educated layman; DNA has become a household word. While the stories presented often are quite accurate, they must be digested by readers whose background in biology is usually sketchy. Even those previously exposed to biology usually learned the subject in a conventional way that usually had little relevance to their gaining an understanding of human biology.

The media outdo each other in presenting lurid stories likely to titillate the jaded appetites of their clientele. The results often are a preoccupation with artifical fertilization, cloning, man-primate chimeras, creation of man to genetic specifications, and other farfetched consequences of the new biology. It is sometimes implied that further advances in biology *must* lead to the universal application of these methods accompanied by an abandonment of conventional reproductive methods and a lowering of the value of human life in general. In his *Brave New World*, Huxley (1) described a future society which practiced cloning and artificial fertilization of individuals preassigned to castes stratified by intellectual ability. Orwell portrayed a totalitarian state in his 1984 (2). Thoughtful human beings rightfully become frightened when these developments are painted as the ways of the future.

Science, 1974, Vol. 185, Issue 4152, pp. 653-663. Copyright 1974 by the American Association for the Advancement of Science.

Comparisons have been made between the current state of biology and the state of nuclear physics before the atomic bomb. It is hoped that by intensive discussions of the possible consequences of the new biology, mankind will be better prepared for the coming of the "biological age" than it was for the "nuclear age." Professionals outside of biology and medicine have become interested in these issues. Lawyers, sociologists, philosophers, and theologians have joined biologists, physicians, and geneticists to discuss the current scene and how to approach the future (3). A new field—bioethics—is being born (4). While there is no dearth of literature in this new field, much of it is somewhat unrealistic.

Research and methods of management of birth defects are intimately tied up with modern biological techniques which, according to the "gloom and doom" prognosticators (5) will lead to the Brave New World. Most researchers in the biomedical sciences and practitioners of medicine have been less pessimistic than many of our confreres in the humanities, social sciences, and theology. In general, those trained in biology and medicine have taken a more pragmatic, but possibly a more short-sighted, view of these new developments. Problems of genetic counseling, intrauterine diagnosis, and screening are with us now and raise a variety of ethical issues quite different from the sensational ones drummed up by some of the mass media.

Biologic Origins of Ethics

Evidence for man having evolved from lower forms of life comes from many areas of biology, including protein chemistry—a field in which extensive studies have revealed similarities in the amino acid sequences of proteins from related species (6). While many details remain unknown, the grand design of biologic structure and function in plants and animals, including man, admits to no other explanation than that of evolution. Man therefore is another link in a chain which unites all life on this planet. Studies of proteins have indicated that man and his closest nonhuman relative—the chimpanzee—differ from each other by no more than do subspecies of mice or sibling species of fruit flies (7). Yet, man differs from all other animals, including the most intelligent chimpanzee, by his ability to use complicated oral and written languages and to conceptualize abstract thoughts. With these unique endowments, our species can create cultures and technologies. We can know our past and worry about our future. We no longer need be subject to blind external forces but can manipulate the environment and eventually may be able to manipulate our genes. Thus, unlike any other species, we may be able to interfere with our biologic evolution. It is most remarkable that the human brain had already reached this supreme position at the dawn of prehistory. The biological substrate that later created the philosophies of Plato and Spinoza, the religions of Jesus and Buddha, the poetry of Shakespeare, Molière, and Goethe, as well as modern science, may have been in existence about 50,000 years ago. There is little evidence that our brains have changed much during this period. Our ancestors some 2000 years ago were certainly similar to us (8).

The building of ethical systems by man may be considered a unique property of the human brain. No other species is known by us to have ethical systems. Just as the human brain gives man his unique language capacity, enabling him to learn to speak Chinese or English, so does the brain give man his "ethical capacity," allowing him to express his values in a variety of ethical systems. A biological substrate for cooperativity and altruism may have developed by natural selection (9). Lone hunters were less likely to survive than those who cooperated with each other. Thus, human "goodness" and behavior considered ethical by many societies probably are evolutionary acquisitions of man and require fostering. Alleviation of suffering, freedom from want, neighborly love, and peace are attributes practically all modern societies would aspire to. Unfortunately, man's altruistic instincts are often overpowered by his aggression. To curb this tendency without dogma and rigid rules is a difficult task facing present-day societies. The human brain is unlikely to change biologically in the foreseeable future. It is also unlikely that man will become extinct. Mass starvation and nuclear war may decimate human populations, but some men are likely to prevail and as long as records of cultural achievements remained in some libraries, high technological achievements would be possible in a few generations.

An ethical system that bases its premises on absolute pronouncements will not usually be acceptable to those who view human nature by evolutionary criteria. New knowledge and new ways of coping with nature offer new and different challenges which the past cannot necessarily help us with. Many persons feel that the consequences, immediate and remote, of a given act should be the sole criterion for judging whether the act is good or bad. Such an ethical system knows no absolutes, no black and white, no a priori do's and don't's, but must laboriously draw up a balance sheet of all the consequences of man's acts (10). Most of us are philosophic utilitarians; that is, we want to do the most good for the largest number. We want this goal achieved by consensus rather than by edict, and we value freedom of action. How free we really are, however, is not entirely clear. Data from such different fields as behavioral genetics and Skinnerian psychology (*11*) raise questions about our cherished beliefs of freedom of action, and until considerably more work has been done in human neurobiology, neurogenetics, and behavioral psychology, we will be unable to settle how open our choices really are.

Fried, a legal scholar with philosophical inclinations, has pointed out that we need a "philosophical anthropology"—a new system that would attempt to harmonize the scientific view of man with existing or new codes of ethics (*12*). He states that existing ethical systems are inadequate and that a consequentialist, situational ethics would also be unsatisfactory. Yet, where is this "philosophical anthropology" to come from? Biology itself cannot provide it, and all philosophical systems are relative and not absolute. In a search for a unifying philosophy of man's existence, Monod (*13*) suggested an "ethic of knowledge" to replace existing beliefs; objective search after the truth and after the truth alone would be the cornerstone of this ethical system. This code would omit all emotional, poetic, and aesthetic human aspirations. It is unlikely that such an austere system would appeal to most poeple

Where do we go from here? I fully agree with Sinsheimer (*14*), a molecular biologist, who said that the enormity of our ethical problems should not paralyze us into inaction. Our recent triumphs in using the brain to help us in our understanding of ourselves and of the universe should not be the terminus but the beginning of a new era of man's life on this planet and even elsewhere. Nevertheless, I urge caution in our applications of current technology. Modern science has been around for only 200 years in man's evolutionary history; biology has been revolutionized only during the last 20 years. We know relatively little about most of human biology, particularly human genetics. Thus the genetic regulation of human behavior and the genetic determinants of normal traits as well as of common diseases and birth defects are largely unknown (*15*). Intensive research on these topics must be conducted and the underlying basic phenomena must be discovered before we attempt to apply genetic knowledge on a grand scale. Yet paradoxical forces exist that tend to spur us to action. Poeple clamor for the fruits of research to be brought from the laboratory into the public domain. Public funds are spent and, for financial support to continue, practical applications are expected in the near future. As a result, premature applications are likely to be attempted. The public wants cures and prevention of disease; yet, for some of the most serious problems the basic knoweldge that would enable us to "deliver the goods" is lacking.

The task of human biologists and physicians is to understand the biology of man and to apply research in a humane and cautious way, with respect for the individual human being. It is likely that in so doing, boundaries will be crossed that were previously considered absolute. The nature of man is to explore and to experiment; to stop exploration and experimentation at this juncture would be to act against those attributes which make us most human.

Surgically Treatable Birth Defects

Many common birth defects, such as a cleft palate, pyloric stenosis, several types of congenital heart disease, and retinoblastoma, can be treated by conventional surgical techniques. Most patients with these disorders are usually cured and made able to lead a normal life; untreated patients with these diseases, except for cleft palate, often die. Developed societies have a sufficient number of surgeons to take care of these defects, and societies now undergoing development include the provision of such surgical treatments in their long-term plans.

All these conditions (with the exception of most unilateral retinoblastomas) have a genetic etiology. The frequencies of congenital heart disease and of pyloric stenosis are 3 percent and 5 percent, respectively, among children of patients, a relative increase of 5 times and 20 times over the frequencies in the general population (16). In the past, patients with cleft palate often remained unmarried for cosmetic reasons. Surgically treated patients now have children and approximately 4 percent of these children are affected. Congenital heart disease, pyloric stenosis, and cleft palate are polygenic conditions that can be expected to double their frequencies in about 20 generations, or 500 years (16). Bilateral retinoblastoma is a dominant trait and will double in frequency from 1 in 70,000 to 1 in 35,000 in a single generation. From that time on, its frequency will slightly increase because new cases will be added by mutation.

Despite these findings, the dysgenic effects of modern medicine have been exaggerated because natural selection before birth, in the form of spontaneous abortions and genetic sterility, still occurs and has not been changed significantly by modern medical practices. What should be done? To discontinue life-saving treatments because of future increases of patients appears unthinkable. Genetic counseling to advise treated patients to have fewer children is highly debatable, and is unlikely to have much success if a person feels that his own single surgical operation was no great problem. Why should his children not undergo such operations? To force parents to be sterilized is repugnant and makes little sense.

Various positive and negative financial incentives in the form of bonuses or taxation have been suggested to discourage high-risk patients from having children, and it is conceivable that future societies might initiate such practices. However, a doubling in the frequency of these conditions is unlikely to place a strain on healthy societies. Industrialized societies require fewer persons engaged in production and must provide more jobs in service industries. A larger number of surgeons and other health care personnel required to take care of more sick people fits modern economic trends. Far from being a serious societal problem, the increase in treatable genetic disease could easily be managed. Furthermore, in the time that it would take for the frequencies of these conditions to double, we might be able to develop methods for diagnosing them in utero or for preventing them. Any public action at this time therefore seems inappropriate.

Medically Treatable Birth Defects

The problems associated with medically treatable genetic diseases are similar to those associated with diseases that can be treated surgically. Restoration of full health and vigor leads to fertility and transmission of previously harmful genes. The problem is illustrated by hemophilia, a condition that can now be controlled successfully by frequent injections of antihemophilic globulin in the form of a special cryoprecipitate. This form of treatment, however, is not as simple a matter as, for example, a single operation at birth for pyloric stenosis. Because the cryoprecipitate is expensive, it costs many thousands of dollars per year to keep a patient well. With normal morality and fertility, the frequency of the disease will double in four generations and triple in ten generations (16)—an increase from the current frequency of 1 in 14,000 to 1 in 7,000 in 100 years. Although we will need additional funds to treat these patients, considering the many other problems of cost in our society, this type of expenditure should not alarm us. Not only is the cost of cryoprecipitate likely to be reduced, but new ways to prevent transmission of the disease might be found. In any case, with education and counseling, many hemophiliacs might opt for a reproductive alternative that would avoid gene transmission (17).

Medically Preventable Birth Defects

Rational plans for the prevention of most birth defects will remain unavailable as long as we do not understand the etiology of these diseases. Preventive measures based on rational understanding of a disease constitute what Lewis Thomas (18) has called "high technology" of disease control. Immunization against rubella to prevent fetal birth defects, and the administration of Rhesus (Rh) antibodies to prevent Rh hemolytic disease would be examples of such "high technology." The only ethical problems associated with these two diseases concern the accessibility of treatments for all women at risk. Rh hemolytic disease might be entirely eradicated if all Rh-negative women were to receive Rh antibodies following pregnancies or abortion. It is conceivable that a significant proportion of birth defects may not be explainable by either genetic or environmental factors, or by genetic-environmental interaction. The complex dynamics of early embryonic organ formation may allow errors on a strictly random basis even after all the genetic and environmental factors affecting development have been fully elucidated. Thus, identical twins share identical heredity and a very similar intrauterine environment. Yet the frequent expression of a birth defect in only one of a pair of identical twins, when other data clearly point to a genetic etiology of that defect, suggests that random factors may play a part in its expression. The cause of the defect therefore could be chance errors that will defy logical explanation. Primary prevention of birth defects based on etiologic understanding is therefore likely to be difficult. For the prevention of most defects we can only recommend that pregnancy be avoided too early or too late in life and that exposure to drugs, chemicals, x-rays, cigarettes, and infection be avoided.

Special Problems of Transplantation

The use of organ transplants for the treatment of birth defects is becoming frequent. Kidneys, for example, are transplanted because of renal failure or as a source of a missing enzyme in conditions such as Fabry's disease (*19*). Bone marrow transplantation has been used in some cases of immune deficiency disease and may have a future in the treatment of hereditary hemoglobinopathies. The distinguished biologist Burnet (*20*) has criticized the use of complex operations, including transplants, in the treatment of disease. He said in 1971: "The application of science to treat any child with a genetic, metabolic, or immunologic anomaly which is potentially lethal, at the best level possible in the light of current knowledge will always be extravagantly expensive. It will require a small full-time team of biochemists and technicians and will subject the child to constant examinations, blood tests, and injections. In most cases such detailed control must be continued throughout life. It is a brutal fact that except in a research center in an affluent country and only when an individual scientist has a professional reputation to make or to maintain in the relevant field, such children can rarely receive empirical care and will die at the first crisis in their condition."

Considerable research effort is spent on transplantation. It is possible that better methods of dealing with the problems of graft rejection will be found so that transplantation might become a simple procedure. In the meantime the ethical problem raised by Burnet must be faced. However, current treatments of the type Burnet refers to must clearly be labeled as research and not as generally applicable therapy. Simpler approaches, however, may result from such investigations which may do away with the complex machinery now required.

Many of the problems that arise with organ transplantation are related to the fact that sibs are usually the best matches for transplants. For example, there are some physical risks in being a kidney donor, since one reserve kidney is removed for the rest of one's life. So should parents be able to volunteer one of their children as a kidney donor for their other sick child? At what age should a child have the right to refuse to become a donor? For what reasons should he be allowed to exercise this right? What are the psychological consequences of refusal and of knowledge that the affected sib will certainly die without the donation? While a pragmatic view would suggest that parents should be able to volunteer their children as donors to siblings, it is evident that new legal safeguards need to be drawn up. We may hope that the use of organs from cadavers, matched to the donor by computer systems covering wide geographic areas, will ultimately obviate the need for sib donors.

Partially Curable Birth Defects

The treatment of spina bifida—failure of the spinal canal to close nomally—illustrates how modern surgical techniques may cause serious ethical problems (21). Many children born with spina bifida and related disorders die unless they are given early neurosurgical treatment. However, a large porportion of children receiving such treatment remain paralyzed, or mentally retarded, or both, and require continual complex medical care. When they reach adulthood, many of them do not procreate; thus, they have little effect on the long-term genetic composition of the population. The magnitude of the problem is illustrated by the situation in South Wales, United Kingdom, where over the past few years the number of families having to deal with such patients has quadrupled from an initial frequency of 1 in 1500.

The principal ethical problem associated with partially curable defects concerns the rescuing of individuals from certain death whose quality of life will be seriously compromised. The general consensus in recent years has been to institute "triage" systems so the surgical treatment is withheld from those who, by empirical criteria, will later on be seriously retarded or paralyzed. Heroic treatments to save their lives are not instituted. The decisions about who should be operated on are generally made by teams of neurosurgeons and pediatricians, usually in consultation with the families of patients. The decision to withhold surgery is made without consultation of the public at large. While involvement by the public in establishing guidelines in such decisions is desirable, direct involvement has not worked too well, for example, in establishing priorties for renal dialysis in adults with kidney failure. Triage systems for these types of birth defects could be abused by governments bringing various pressures to bear on the professionals to extend their criteria for withholding of treatment. Free and open discussions of all such issues may be one of the best safeguards against perversion of biological and medical practices.

Tests to diagnose spina bifida and related disorders in utero have become available, and even blood tests for these conditions are on the horizon (22). Abortion of affected fetuses following intrauterine diagnosis is already possible if the mother is known to be at high risk because of a previously affected child. Complete prevention of this disorder would require amniocentesis of all pregnancies—a yet impractical suggestion—but such total prevention would be possible if a blood test were available.

Screening for Phenylketonuria and Related Conditions

Screening at birth for certain diseases that can be treated early in life has become accepted medical practice in many countries. Phenylketonuria (PKU) is one such disease that can be detected at birth by simple blood tests. The frequency of this condition, which leads to marked mental retardation in untreated patients, is around 1 in 10,000 births. Treatment consists of restricting the dietary intake of phenylalanine, and current results suggest that treated children with classic PKU develop normally or almost normally. However, a positive screening test does not necessarily mean that the baby has PKU, because variants exist which may not cause mental retardation. A team of experienced biochemists and pediatricians is therefore required to follow up children showing positive tests to ensure that treatment is administered properly.

Although it appears that in children with PKU the diet can be stopped at around 6 years of age, when high concentrations of phenylalanine no longer injure the brain (23), girls with PKU need to be informed about the risk of their children being mentally retarded. The high concentration of phenylalanine in the maternal blood crosses over to the fetus and causes severe mental retardation in all children of such women. Presumably, reinstitution of the phenylalanine-free diet during pregnancy would prevent this problem.

An affluent society can afford to screen 10,000 normal children to find the single child with PKU who can be protected against mental retardation by appropriate diet. Sponsors of screening plans have pointed out that in addition to the prevention of tragedy in a family, the cost of the screening program is significantly less than the cost of caring for a single patient with PKU over his lifetime. However, since there is no reason to believe that a decline in numbers of patients with PKU (1 percent of patients institutionalized with mental retardation) would have much effect on the total budgets of hospitals for the mentally retarded, it would be better to use cost-effectiveness analysis only for those disease categories that would reduce total patient load by a much higher proportion.

There are other inborn errors of metabolism, such as maple syrup urine disease, homocystinuria, and galactosemia, that can also be screened for at birth, but each of these conditions is extremely rare. Homocystinuria is probably the most common, with an incidence of 1 in 160,000 births (24). Screening for such rare diseases which often cannot even be effectively treated is difficult to justify. However, once testing has been established for PKU it may be relatively inexpensive to add tests for other more rare diseases to the screening system. Obviously, large-scale pilot investigations with careful long-term follow-up studies will have to be conducted, and all the psychological effects and the economic costs of following up false positives will have to be taken into consideration, before a final assessment can be made.

Screening for genetic diseases often causes racial problems. For example, PKU is so rare in the black population that the screening of black children for this condition can be seriously questioned. Yet some observers would object to a scheme whereby blacks were omitted from screening, on grounds of its being discriminatory. Should we test for Tay-Sachs carriers in non-Jews? For sickle cell trait carriers in whites? My recommendation would be to limit testing to relatively small populations in which the particular disease occurs at a high frequency, even if the disease occurs very occasionally in the more populous racial groups.

Screening for PKU is a problem of resource allocation, particularly in the developing countries. Many countries lack the required logistic system to reach and follow up each child. To screen for rare genetic diseases at a time when malnutrition and infectious diseases are the principal problems would be a waste of resources. The continuing and ever-growing contrast between health care in the Western world, the Soviet Union, Japan, and China on the one hand, and most other areas of our planet on the other, is an ugly monument to the failure of the human species to cooperate. The ethical problems related to birth defects are small when compared with the colossal health problems of the developing countries. How to bridge this gap is a moral problem which too many of us shrug off in helplessness. The population issue must be faced in this context. Reduction of nutritional and infectious disease in the absence of birth control raises worse problems.

Genetic Counseling

Genetic counselors usually are physicians with training in medical genetics who first make an accurate diagnosis of the disease in question, and then provide their patients with information about the natural history of genetic diseases, about the risks of these diseases occurring in offspring, and about the available alternatives to bearing affected children. The aim of such counseling is to enable a couple or a person to make rational decisions about whether or not to reproduce. Although at least one follow-up study has suggested that those who have been counseled get a good grasp of the meaning of risk and avoid reproducing if the risks of their having affected children are high (25), some data (26) have indicated that the meaning of genetic risk may not always be well understood.

Most counselors consider their work to be little different from any other medical practice; they put the interests of the patient and his family before the interests of society and the state, and pursue medical, not eugenic, objectives. Untoward effects on society may be pointed out, but most counselors do not attempt to give advice based on considerations of the gene pool.

Genetic counseling has thus, traditionally, been nondirective. It is usually maintained that every family situation is different and that the meaning of a given risk varies from family to family, so that in some cases even high recurrence risks may justify a future pregnancy. Some critics (27) have suggested that families expect more definite advice than is often provided, saying that because a genetic counselor understands the total impact of the disease and the real meaning of risks better than does the family, he should advise what he or she thinks would be the best course of action. Until better studies have been made of these matters, it will be impossible to make any firm conclusions. In the meantime, depending on the assessment of the counseling problem, many experienced counselors are usually nondirective, but may occasionally alter their approach.

There are also some broad ethical issues associated with genetic counseling. The motivation of a couple who by their own initiative seek counseling is usually different from that of a couple referred to counseling by a physician or some other interested party. Those who seek genetic counseling may be better educated than those who do not and they may, consequently, obtain a better understanding of the risks for their future offspring. With an increasing availability of genetic services, more people who may be unaware that a genetic problem exists or who may not be motivated to seek advice, may nevertheless receive genetic counseling. Under such circumstances, the counseling may be "forced" upon persons. Provided that the information given them is nondirective, they will probably not object, but if a counselor advises reproductive restraint, for example, when such advice has not been asked for, the probelm will be more difficult. In countries where private health insurance programs are the rule, it might be possible for the insurance companies involved to withhold benefits from a sick child born to parents who were advised not to reproduce. In countries where health insurance is nationalized, regulations for withholding benefits from certain patients with genetic disease would probably be difficult to administer, and might not be passed for this reason.

Although laws authorizing the sterilization of certain patients have existed in the United States for many years, most of them are no longer applied. The excesses of Nazi Germany in this regard are not many years behind us. Recent newspaper reports of the sterilization of retarded black girls in the southern United States created much furor. Although a logical case can be made for the voluntary sterilization of persons who carry certain harmful genes, who should make the decision for those persons lacking the intellect to decide for themselves? Legal safeguards to prevent possible abuses of existing laws are absolutely necessary and no decision to sterilize a person should be made without the concurrence of representatives of that person's family, the legal and the medical professions, and public representation at large. Laws authorizing enforced sterilization for genetic reasons should be strongly rejected, largely because the rights of couples to make their own decisions, even if this decision might result in the birth of a defective child, must be defended. Improvements in education in human biology, and a greater availability of genetic counseling and related services, should go a long way toward enabling people to make rational decisions about reproduction and toward reducing the numbers of children born with genetically determined illness. The marked change in popular attitudes toward abortion in many societies is a good example of how attitudes regarding reproductive practices can alter rapidly.

Other ethical problems may arise when genetic counseling is extended to family investigations. Some genetic diseases may be delayed in the onset of symptoms and their carriers may not know that they are affected. Identification of a clinically affected patient allows the performance of diagnostic tests of the relatives at risk. Following diagnosis of a disease in its early stages life-saving treatment may be initiated. There is little question that case-finding among relatives is strongly indicated when treatment and prevention of the disease is possible, for example, in hereditary polyposis of the colon, Wilson's disease, and porphyria. How far should the physician or medical geneticist go in order to trace all persons at risk? Should health departments get involved to ensure case-finding among scattered families? Who should be responsible for checking that everyone at risk has been examined?

More problems arise if a disease is clearly genetic in origin but no definite treatment is available (for example, Huntington's chorea). Should one attempt to detect those who are affected before they are clinically ill? Would most persons want to know many years before symptoms develop that they will die prematurely of an incurable disease? If a person were told that he had a high probability of developing such a disease in his middle years, he might decide not to have children. What can be used as guidelines? The relatives may be completely unaware of the risks. The very communication of the problem might create anxiety in a person even if he or she decided not to pursue the matter. Should we insist that relatives at risk be given the relevant information? While no generally applicable rules can be made, many observers point out that some information is better than none, and that the relatives have a right to know. The withholding of information is considered a form of medical paternalism. Nevertheless, some physicians occasionally decide not to pursue investigations to detect a genetic disease in family members when nothing can be done to prevent or cure the disease. Often the patient and his immediate family can provide assistance or guidance concerning the potential interest of other relatives in genetic counseling. We need many more data on these matters.

Intrauterine Diagnosis

The development of intrauterine diagnostic techniques, such as amnio-centesis, for the detection of chromosomal errors, X-linked diseases, and certain inborn errors of metabolism, is revolutionizing genetic counseling (28). New sonographic and optical methods are being explored and may widen the scope of intrauterine diagnosis for other conditions. However, most problems requiring genetic counseling cannot yet be approached by intrauterine diagnosis. Researchers in this field foresee a time in the future when amniocentesis may be used routinely in the monitoring of most, if not all, pregnancies. Before this comes about, however, a variety of technical and logistical problems will have to be solved. More diseases need to be diagnosed and the absolute safety of the mother and fetus will have to be ensured. While it is difficult to prophesy, wide use of this procedure appears more likely than some of the more futuristic biological schemes under discussion. Unless it is used for every pregnancy, intrauterine diagnosis will have little impact on the population frequency of most birth defects (29).

When a fetus is found by intrauterine techniques to be genetically defective, the parents usually choose to abort it. Abortion causes serious ethical problems to many people for religious or personal reasons, although, first in Japan and more recently in the United States, there has been a rapid change in public acceptance of this procedure. The abortion of a fetus affected with a devastating disease such as Tay-Sachs or mongolism is accepted by many individuals who would oppose abortion for reasons of family limitation or convenience. More difficult problems will arise, however, as milder genetic defects are diagnosed as an unexpected finding following amniocentesis for indications of more harmful diseases. Should an abortion be performed for Klinefelter's (XXY) or Turner's (XO) syndromes? What about cleft palate where a single operation would cure the affected child? Difficult decisions will have to be made about the normality or abnormality of a fetus, because any fetus not considered up to "standard" might be rejected. The problem would became particularly acute should intrauterine diagnosis and abortion become simplified and more widely available. If it becomes possible to diagnose, and thus to abort, defective fetuses at an earlier stage in development than is now possible, many people might choose abortions who would hesitate to undergo this procedure during the early portion of the second trimester of gestation as is now required.

Scenarios have been considered in which the state enforces abortions to save money that would otherwise be spent on the care of persons with severe birth defects. Such a step seems unlikely. It is more probable that most people would volutarily seek this method of avoiding birth defects. Some observers have suggested that the widespread acceptance of intrauterine diagnosis by many couples might lead to public rejection of children with preventable birth defects who could have been aborted. This development is also improbable: attitudes of the public and of medical personnel toward patients with cancer of the lung, which could have been prevented by their not smoking, is no different from attitudes toward patients with cancer of the colon, which we do not know how to prevent.

Many physicians refuse amniocentesis to pregnant mothers who say that they will not undergo abortion in the event of the fetus being found defective. They say that for these mothers, the early diagnosis of untreatable disease in the fetus would be harmful psychologically and would serve no purpose. In opposition to this viewpoint, one can point out that most amniocenteses give normal results, and that the total happiness generated in families receiving such results outweighs the anguish of the rare couple who know that they will have an affected child but choose not to abort. It is therefore difficult to establish absolute values regarding who should and who should not be given tests that are available.

The possible dysgenic consequences of selective abortion have been considered in detail elsewhere (29). While these practices will cause some increase in the numbers of deleterious genes, few serious long-term problems are likely to arise.

The most serious question concerning the ethics of widespread abortion to prevent the birth of genetically defective children is based on the following reasoning: Why go to all the trouble and expense of doing intrauterine tests that might harm the fetus if inspection of the infant and diagnostic tests at birth would be much easier? An infant with serious birth defects could be "terminated" at that time (30). Proponents of this viewpoint suggest that a newborn baby should not be considered legally "human" until certain standards of normality have been assured, pointing out that passive infanticide, that is, the withholding of treatment, has always been practiced with severe birth defects. There are awesome implications in these arguments. Most societies differentiate between life in the womb and life after birth. Each month of pregnancy allows for the development of emotional bonds, particularly between the mother and her infant. Perhaps because of recognizing these bonds, most societies in the 20th century have rejected infanticide and place great value on human life after birth. To practice active infanticide for medical purposes to me appears regressive and loathsome, and in effect would officially sanction already existing trends toward the blunting of human sensitivity. The next step, logically, might be the extension of such practices to so-called "mercy killings" at all ages of life, starting with the aged and incurably ill. The experiences of Nazi Germany only 30 years ago show that such practices, which were followed by genocide of almost half of the world's Jewish population, can become a reality.

It is sometimes said that selective abortion after intrauterine diagnosis is an interim measure, and that in the future it will be possible to treat birth defects and genetic diseases either pre- or postnatally. This view is probably unrealistic. Efficacious treatment for a complex defect such as Down's syndrome and similar structural defects is difficult to imagine. Many types of existing and future postnatal therapies cause a certain amount of suffering in the child. Prenatal therapy applied to the fetus may be dangerous to the mother also. Therefore, even when effective treatments for more birth defects have been developed, many parents will probably prefer a safe abortion with the assurance that their next child will not be affected with the disorder for which selective abortion was performed. This means that abortion for genetic defects discovered by intrauterine diagnosis is here to stay for a long time.

The control of common recessive diseases, such as cystic fibrosis, is most likely to be achieved by detecting heterozygous carriers before or after marriage or mating and by developing methods that will enable physicians to differentiate between normal, heterozygous, and affected fetuses by intrauterine techniques. Carriers would be informed of the 25 percent probability of their offspring being affected if they mated with a carrier of the same disease, and diagnosis could be made in utero, with the mother having the choice of aborting an affected fetus.

Such an approach, if applied by a large fraction of the population, would reduce the numbers of children born with such recessive genetic diseases, and therefore might receive high priority in the allocation of funds for medical research. As a consequence, more basic investigations of this and similar diseases might be deemphasized, or abandoned, in favor of developments of methods leading to the intrauterine diagnostic approach. It is therefore conceivable that treatment of recessive diseases based on a fundamental causative understanding might not be developed because of lack of research efforts. While the discovery of good screening methods and intrauterine tests requires a certain amount of basic understanding, it is clear that the goal of intrauterine diagnosis is more limited and requires fewer total resources than more comprehensive research.

Sex Choice

Determination of the sex of a fetus is already feasible with amniocentesis, and this makes possible sex choice by selective abortion. Since the procedures are somewhat novel, and since a second trimester abortion is required, this technique is rarely used except to detect and abort male fetuses affected with genetic diseases which are linked to the X chromosome, such as hemophilia and the Duchenne type of muscular dystrophy. The procedure is usually refused to couples who desire a child of a certain sex after they have had several children of one sex only. If prenatal amniocentesis becomes a routine procedure, however, sex choice will probably be practiced more often.

While abortion as a means of sex choice may be objectionable, other more acceptable procedures by which to choose the sex of a child may soon be discovered. For example, it might become possible to separate X from Y sperms, in which case sex choice by using the husband's X or Y sperms for insemination would be a simple way of having children of the desired sex. A sociologist has pointed out that if sex choice were widely practiced, more males would be selected than females and, because of this, there would be significant long-term effects on society, such as an increase in homosexuality (*31*). The social effects of a preference for male children would be delayed for almost a generation, however, and it is of interest in this regard that the state of Alaska already has an excess of males, but has not encountered serious societal dislocations.

Significant changes in sex ratio could probably be avoided if the composition of the population were carefully monitored, so that any deviation from an acceptable ratio could be brought to the attention of the public. Widely disseminated discussions regarding possible untoward consequences might then change preferences in sex selection of children.

The recent rapid change in styles of family size indicates that swift alterations in reproductive practices do occur. Thus, there is no indication that research on sex choice should be placed under rigid control. In fact, such research should be encouraged, since the discovery of a simple method for choosing the sex of children would allow ideal family planning.

Population Screening for Genetic Reasons

Screening for diseases, such as PKU, which are potentially treatable or preventable by medical or surgical methods, raises fewer problems than screening for conditions for which patients require either conventional genetic counseling about recurrence risks or intrauterine diagnosis following genetic advice. There are several recessive diseases, such as sickle cell anemia, thalassemia major, Tay-Sachs disease, and cystic fibrosis, that are either very difficult to treat or cannot be treated effectively. Each of these conditions is relatively frequent in a certain ethnic group; the conditions range in frequency from 1 in 100 for sickle cell anemia in certain populations in Africa to 1 in 4000 for Tay-Sachs disease in Ashkenazi Jews. Such frequency figures indicate that a significant fraction (3 to 25 percent) of the respective populations are heterozygous carriers for the relevant genes. Tests for detecting carriers of these diseases already exist (except for cystic fibrosis). When carriers receive counseling, they are informed of the 25 percent chance of their children being affected if they marry a carrier of the same gene. In Tay-Sachs disease, intrauterine diagnosis and selective abortion of affected fetuses is already possible. To be most effective, testing procedures should be initiated prospectively, that is, before a child with the disease is ever born. Retrospective counseling following the birth of an affected child is not an effective means of disease prevention since only 12.5 to 25 percent of cases can be prevented in this way (29). Some geneticists believe that even in the absence of intrauterine diagnosis, population screening followed by genetic counseling of all carriers would cause a reduction in disease frequency because of reproductive restraint among married carriers or appropriate mating choice among those not yet married.

In practice, the widespread screening for sickling in the United States probably has had several untoward consequences (32). Many screening programs were set up without counseling components, and many carriers of the harmless sickle cell trait, because they were not informed otherwise, came to believe that they had a mild form of, or a tendency to, sickle cell anemia, Social stigmatization, occupational discrimination, uprating of insurance premiums, and psychologic invalidism of sickle cell trait carriers were among the results of these programs. In addition, there was a lessened choice of marriage partners for the many people who mistakenly believed that a sickle cell trait carrier was a less desirable mate. In some instances, when a child with a positive sickling test was found to have two nonsickling parents, the illegitimacy thus detected became known to the legal father.

These well-meaning screening programs therefore produced serious problems because the social consequences to a person being identified as a carrier were not taken into consideration (*33*). Before anyone is asked to give consent for screening, they should be fully informed of all the possible social, as well as medical, consequences of being diagnosed as a carrier. Certainly, before programs for screening the total populations at risk are developed, there should be an extensive assessment of existing practices. Many, but not all, of the problems in sickle cell screening apply to the screening of other diseases of this type.

Much of the misunderstanding about sickle cell anemia and other recessive diseases would certainly be eliminated if the entire population at risk received special educational programs during their early years. Genetic counseling of trait carriers alone would not be satisfactory because the total population at risk needs to be informed.

A better long-term solution to the problem of sickle cell anemia and other hemoglobinopathies, in my view, would be the development of techniques for diagnosing them in utero as are already available for Tay-Sach's disease. The carrier status of a potential mate would then be less important since intrauterine tests could be offered to all couples where both partners were carriers, and affected fetuses could be aborted if desired by the parents. Although this approach has raised cries of "genocide" among some black leaders, programs of this kind for Tay-Sachs disease are in operation in some Jewish communities (*34*). The approach used in these programs is an attempt to make all members of the Jewish community aware of the disease and of its frequency in the Jewish population, and to encourage all members to be tested. An end result similar to that obtained by screening the total population at risk could be obtained if obstetricians tested all pregnant Jewish woman (*35*) and arranged for the testing of husbands only if their wives showed positive tests. More medically oriented schemes of this type have the advantage of arriving at the same results without alarming the whole community. On the other hand, in the United States the community approach appears more practicable at this time than enlisting the cooperation of obstetricians and general practitioners who attend the pregnancies of Jewish women.

A program that omits community participation runs against the current trends that aim at maximum dialogue between experts and the public. Nevertheless, complex issues of this kind are understood with difficulty by many people and therefore will cause unnecessary anxiety. Physicians are not required to inform their patients about all possible medical risks of a given procedure, for if they did so, every simple intervention might cause much anxiety. In a recent court ruling, it was stated that untoward risks of a medical or surgical procedure that carry a risk of 1 percent or less need not be discussed with patients. An analogous rule for genetic diseases might be worked out. Thus, genetic diseases that affect fewer than a small fraction of the population might best be dealt with medically without extensive community involvement.

As soon as the absolute safety of intrauterine diagnosis is established, it would be prudent to initiate the screening of all pregnant women older than about 38 years for fetuses affected with Down's syndrome (mongolism). All physicians, regardless of their attitudes toward abortion, should know about the procedures and should fully inform appropriate patients about the possibility of their giving brith to affected children and about the alternatives available. Fortunately, with more effective and more widespread family planning, there will be fewer pregnancies among women of relatively advanced maternal age, and consequently fewer cases of Down's syndrome.

Problems in Early Detection of
Genetic Disease of Late Onset

In the future it might be possible to detect early in life, even at birth, a variety of diseases that may cause medical problems later in life. In this category are the hyperlipoproteinemias which predispose affected persons to myocardial infarction in middle age (36). Although we have no proof yet that drugs and dietary manipulations defer the onset of coronary disease if instituted early, such an outcome is likely. We should, therefore, consider some of the problems that might have to be faced in the future. For example, what would be the reaction of parents who were told that their hyperlipidemic child had a 50 percent chance of having a heart attack at age 50 years? Would this be enough of a risk to make them change the family diet, or administer a drug all through childhood? Would they be willing to educate the child to a life style that would reduce the probability of his having a heart attack? Would it be child neglect if the parents refused to use a medical or dietary regimen that would help the child 50 years later? Would society be able to ensure in some manner that children would be provided with the environment their genotype required for optimum health?

Particularly difficult problems would be encountered if it became possible to identify future psychiatric disease. We already know that a person with an XXY chromosomal constitution (Klinefelter's syndrome) has an increased risk of suffering mental retardation and of minor sociopathy, but we know of no way to reduce these risks. What should the parents of an XXY child be told? If amniocentesis were to be applied universally, the identification of XXY would probably lead to abortion of many such fetuses. The data regarding the XYY chromosome pattern are still too confused (37) to be certain about the risks of criminal or antisocial behavior. It is certain, however, that many parents would choose abortion of an XYY fetus if the risk of its showing such behavior were significantly increased over that of the general population.

The detection of individuals predisposed to schizophrenia could create serious problems if we did not also find a way to prevent the manifestations of the disease. Would it not be tragic for parents to know that their newborn child would develop a serious crippling mental disease at age 20 years? While research on the genetics of schizophrenia and on the effects of the environment on the manifestation of the disease is proceeding, many families participating in this research may obtain information about their children that they probably would rather not know. If a major gene for schizophrenia could be identified, it is likely that tests for this gene could be done in utero and many parents might decide to abort an affected fetus. Thus, it is clear from developments in many genetic diseases that intrauterine diagnosis followed by selective abortion will have wide applicability.

Artificial Insemination

Artificial insemination with donor sperm (AID) in cases of male infertility has been practiced for many years. The practice is handled by a few physicians who usually select donors to match the husband's general appearance and background. Genetic investigation of the donors by history or laboratory tests is not usually done, and the legal status of children born after AID has not been well defined. The use of donor sperm if both members of a couple carry the same recessive gene can prevent the birth of a defective child. This practice, however, is not often selected in genetic counseling as an alternative method of reproduction.

In recent years, sperm banks have been formed in several cities of the United States to make it possible for men to leave a specimen at a bank before undergoing vasectomy, so that they can have children if for any reason they choose to do so. While the short-term storage of sperm appears to be safe, the effects of long-term storage have not been fully tested, and research in this area raises problems concerning human experimentation. There are many questions that have not been satisfactorily answered by existing sperm banks and none of the banks, as far as I know, have been licensed by any federal or state agency.

Sperm banks could be used to widen considerably the selection of potential donors for AID, particularly if all donors were subjected to genetic investigations. Such a development would bring us close to Huxley's *Brave New World,* except that the donors would be chosen by the physician in consultation with the couple, rather than by the state. The usual practice of the donor remaining anonymous to the couple receiving AID would probably be continued because it prevents undue psychological attachment by the mother to the donor. Sperm banks could also be used for the storage of sperm by young men who might want to have children later in life, but who want to reduce the risk of mutations that occur at a higher frequency in the sperm of older persons. Persons receiving exposure to radiation or mutagenic chemicals could avoid potential problems by depositing their sperm in a bank before exposure. The use of sperm from outstanding human individuals has been recommended by Muller (*38*) as a method of upgrading the genetic constitution of man. He recommended that the sperm be stored until the "candidate" had died so that there could be general agreement about his social worth. Such a scheme is unlikely to be adopted by most women. Furthermore, we know too little about the genetics of desirable human qualities to be able to forecast the outcome of such a practice.

Fertilization in vitro

There has been much discussion about the problems that would arise if it became possible to produce human "test-tube babies" (*39*) In this procedure, human ova would be removed from a woman and would be fertilized in vitro. After some cell divisions the resulting blastocyst would then be reimplanted in the uterus (*40*) where development would proceed as in normal pregnancies. Such "test-tube babies" can be produced in lower species and only a variety of technical, rather than conceptual, obstacles prevent the application of the procedures to man. In its simplest application, a woman with a blocked fallopian tube could thus become pregnant with her own ova fertilized by her husband's sperm. The ova and sperm could, however, be from any human source, and any woman could serve as the "baby carrier."

Fears have been expressed that a government might use these techniques in schemes to breed its citizens. However, in the absence of knowledge of the genetics and the gene-environment interaction of most normal human traits, directed human breeding is not possible; the results would be no more predictable than they are now when a couple has a child in the conventional way. A moratorium on research in this area, as suggested by the American Medical Association (*41*), would serve no useful purpose because prohibition on research in one country can easily be circumvented by such research being conducted elsewhere. Only a moratorium declared by an international commission would be likely to have results. Pointing out the possible dangers of misuse in a wide variety of forums may be the most effective means of preventing abuse.

The achievement of human fertilization in vitro will raise many problems concerning the safety of the procedure. For example, how could a couple give consent to a procedure that might lead to their child being born with a birth defect? This problem is analogous to many earlier situations where women have used birth control pills or antifertility drugs that might have harmed the fetus. Experimentation with long-stored sperm raises similar difficulties. The prohibition of experimentation of this kind with human beings would effectively stop a wide variety of studies. For example, all work on intrauterine diagnosis would have to stop because we cannot yet be absolutely certain that a fetus subjected to amniocentesis is not harmed in some subtle way. If the use of fertilization in vitro for human beings were preceded by thorough studies of the process in subhuman primates, some of the risks involved might be reduced. Fortunately, the early period of embryonic development appears particularly resistant to birth defects in experimental animals, so that some observers feel that experimentation with nonhuman primates could be dispensed with. Furthermore, any pregnancies in women brought about by fertilization in vitro could be very carefully monitored by chromosomal and other inspection techniques. Even with these precautions, however, some defects might not be discovered and a defective baby might be born. In this context it should be recalled that at least 2 percent of all infants born following conventional pregnancy have severe birth defects.

Provided that the decision to use fertilization in vitro is made voluntarily by the couple wanting a child and, ideally, provided that physicians other than the investigators make the couple fully aware of all the potential dangers, there should be no reason to prevent such a couple from participating in this kind of human research. Sufficient experience might be gained in this manner to ensure its safety. Because a couple who would otherwise be sterile could, by fertilization in vitro, be given a chance of having a normal child of their own, it is difficult to agree with those who suggest that normal procreation is human and fertilization in vitro is inhuman (39). I consider novel reproductive technologies as a more human activity than making babies in the usual way. Thus, reproduction by intercourse in man differs little from sexual reproduction in most animal species. Nevertheless, it is clear that unlike the prevention and treatment of relatively rare genetic diseases which may be considered as medical problems, the various social and ethical issues raised by fertilization in vitro deserve wide discussion.

Embryo Research

If we are to acquire new insights into the biology of man and his birth defects, studies of developmental biology must include research on human embryos. A large number of embryos are aborted in the United States, Japan, and the Scandinavian countries. Some of these embryos, or at least parts of them, are used for studies aimed at understanding mechanisms of development. Since most of the embryos are dead within minutes of being aborted, and since autopsy is a medical tradition, these studies raise few ethical problems. Difficulties arise when embryos removed by abortion procedures are kept alive for research purposes. The use of living embryos facilitates studies of human development and of the effects of physical, chemical, and infectious agents on the embryo. While many individuals might not object to a fetus being kept alive for several hours, they might seriously object to a fetus being kept alive for the purpose of an experiment that might take days or even weeks to complete. Informed consent should undoubtedly be obtained before an embryo is kept alive for research purposes. Presumably, the mother who is to be aborted would be the most appropriate person to give such consent.

There are no compelling medical reasons to attempt ectogenesis, that is, fetal development entirely outside the body. However, a large amount of information that might eventually be of great value in finding methods for the prevention and treatment of birth defects could be gleaned from prolonged studies of fetal development in vitro, particularly if early embryos were used. While many biologists and medical investigators do not consider such studies to be unethical, there is sufficient public criticism of this work that any study of this sort needs the most meticulous scrutiny. However, the outright condemnation of such investigations must be deplored.

Cloning

Cloning of man would involve the creation of a human being who was genetically identical to the donor of a somatic cell nucleus implanted in an enucleated egg. Cloning has been accomplished in amphibians and has been discussed as a possibility for mammals (*42, 43*). Huxley anticipated the process by the Bokhanovski procedure in his *Brave New World*. If cloning of man ever became possible, there is no reason to believe that it would be widely used, even for medical purposes. Cloning as a means of dealing with genetic disease by duplicating either the mother's or the father's genotype is unlikely to be utilized.

The creation of groups of cloned military scientists or brute soldiers in the service of a state bent to conquer the world is a remote possibility. If such an event were to occur, one might presume that other countries would respond by cloning similar groups of individuals. However, simpler ways of subjugating people would probably be more attractive to politicians than cloning, because clones would take as long to develop as any normal human being. Extensive discussions of cloning can be found elsewhere (*42, 43*). Since so many ethical problems of immediate urgency are with us now, problems associated with cloning can be dealt with in due course—if they ever arise.

Gene Therapy: Genetic Engineering

Recent developments in molecular biology allow the synthesis of genes, and the possibility of introducing genes into cells by viral transduction has been raised. As a result, much recent discussion has been devoted to gene therapy (44). More generally, even the possibility of creating human beings to genetic specifications has been raised. Unfortunately, too much has been promised in this field. First of all, only defects in Mendelian traits whose biochemistry is understood (or possibly polygenic traits to which a major gene contributes) could be approached with gene therapy. We know next to nothing about the control mechanisms of mammalian cells. While a gene whose messenger RNA can be isolated can now be manufactured relatively easily, its safe introduction into the nucleus of a specialized cell followed by normal function remains exceedingly problematical. Moreover, each genetic disease presents different problems of gene therapy. While gene therapy of somatic cells appears far away, gene therapy of eggs or sperm, or of gonads, with ultimate genetic cure may never be achieved (45). A group of scientists interested in gene therapy have disclaimed an interest in that aspect of gene therapy which would preserve deterimental genes or maintain them in the population [see (45)].

Clinical investigations of gene therapy by means of viral transduction will raise serious ethical problems because of the possibility of untoward consequences such as cancer. Initially, gene therapy could be tried only for the most severe and lethal diseases, and only after very careful animal experimentation. However, most genetic diseases which are conceptually amenable to gene therapy are individually rare. The more common genetic diseases and birth defects are multifactorial and would not respond to gene therapy unless a major manipulatable gene could be identified. Most normal traits are polygenic, so that the manufacture of a man according to genetic specifications must remain in the realm of science fiction. The genetic manipulation of viruses for the prevention of viral diseases, possibly including cancer, is more likely to be achieved than is the management of genetic diseases by gene therapy. Similarly, genetic engineering of plants to provide more food for hungry man is another exciting possibility. In general, it has become clear that the techniques of intrauterine diagnosis and abortion of defective fetuses will be of much greater importance in the control of birth defects than will gene therapy.

Summary

Recent developments in biology and medicine are raising new problems in the prevention and treatment of birth defects, and in research on these diseases. The problems include immediate issues such as genetic counseling, abortion for birth defects, the withholding of complex treatments from individuals in some situations, screening for genetic and other diseases, artificial insemination, and fertilization in vitro. Other problems, such as the dysgenic effects of modern medicine and the possibilities of cloning and gene therapy, are more remote. Each of these issues should be considered on its own merits and by its immediate and remote consequences rather than by a priori absolute criteria. Ways must be found to deal with these issues in a manner acceptable to most human beings. Open discussions and freedom from coercion are the best guarantees for ultimate success. The ethical human brain is the highest accomplishment of biologic evolution. By harmonizing our scientific, cultural, and ethical capabilities, the potentially achievable results can place us at the threshold of a new era of better health and less human suffering.

References and Notes

1. A. Huxley, *Brave New World* (Harper & Row, New York, 1932).
2. G. Orwell, 1984 (Harcourt, New York, 1949).
3. M. Hamilton, Ed., *The New Genetics and the Future of Man* (Eerdmans, Grand Rapids, Mich., 1972); B. Hilton, D. Callahan, M. Harris, P. Condliffe, B. Berkley, Eds., *Ethical Issues in Human Genetics* (Plenum, New York, 1973).
4. D. Callahan, *Hastings Cent. Stud. No. 1* (1973), p. 66.
5. J. R. Maddox, *The Doomsday Syndrome* (McGraw-Hill, New York, 1972).
6. M. O. Dayhoff, Ed., *Atlas of Protein Sequence and Structure* (National Biomedical Research Foundation, Washington, D.C., 1972), vol. 5.
7. M. C. King and A. C. Wilson, *Genetics 74s* 140 (1973).
8. G. S. Omenn and A. G. Motulsky, in *Genetics, Environment, and Behavior,* L. Ehrman, G. S. Omenn, E. Caspari, Eds. (Academic Press, New York, 1972), chap. 7, p. 131.
9. C. H. Waddington, *The Ethical Animal* (Atheneum, New York, 1961).
10. J. Fletcher, *N. Engl. J. Med.* **285**, 776 (1971); in *The New Genetics and the Future of Man*, M. Hamilton, Ed. (Eerdmans, Grand Rapids, Mich., 1972), chap. 3, p. 78.
11. B. F. Skinner, *Beyond Freedom and Dignity* (Knopf, New York, 1971).
12. C. Fried, in *Ethical Issues in Human Genetics,* B. Hilton, D. Callahan, M. Harris, P. Condliffe, B. Berkley, Eds. (Plenum, New York, 1973), p. 261.
13. J. Monod, *Chance and Necessity* (Random House, New York, 1971).
14. R. L. Sinsheimer, in *Ethical Issues in Human Genetics,* B. Hilton, D. Callahan, M. Harris, P. Condliffe, B. Berkley, Eds. (Plenum, New York, 1973), p. 341.
15. For a discussion of the current state of medical genetics, see A. G. Motulsky, *Am. J. Hum. Genet.* 23, 107 (1971).
16. World Health Organization Scientific Group, *WHO Tech. Rep. Ser. No. 497* (1972).
17. _____, *WHO Tech. Rep. Ser. No. 504* (1972).

18. L. Thomas, *Saturday Review* 55 (No. 52), 52 (1972). Thomas has pointed out that our approaches to most diseases involve "half-way technology" or "no technology." The expensive "half-way technology" of artificial respirators as compared with the inexpensive vaccine (high technology) in the control of poliomyelitis is a classic example of the past. At the present time, more and more resources are used by "half-way technology" modalities of treatment such as coronary care units and artificial kidneys. Thomas calls for more basic research to prevent coronary atherosclerosis and chronic nephritis to achieve "high technology" solutions to these problems rather than sole concentration on the expensive "half-way" measures.

19. D. Bergsma, Ed., *Natl. Found. March Dimes Birth Defects Orig. Art. Ser.* 9, No. 2 (1973).

20. M. Burnet, *Genes, Dreams and Realities* (Basic Books, New York, 1971).

21. H. B. Eckstein, *Br. Med. J.* 2, 284 (1973).

22. O. J. H. Brock and R. D. Sutcliffe, *Lancet* **1972-II**, 197 (1972).

23. N. Holtzmann, personal communication.

24. H. L. Levy, *Adv. Hum. Genet.* 4 (1973).

25. C. O. Carter, K. A. Evans, J. A. F. Roberts, A. R Buck, *Lancet* **1971-I**, 281 (1971).

26. C. O. Leonard, G. A. Chase, B. Childs, *N. Engl. J. Med.* 287, 433 (1972).

27. B. Childs, personal communication.

28. A. Milunsky, *The Prenatal Diagnosis of Hereditary Disorders* (Thomas, Springfield, Ill., 1973).

29. A. G. Motulsky, G. R. Fraser, J. Falsenstein, *Natl. Found. March Dimes Birth Defects Orig. Art. Ser.* 7 (No. 5), 22 (1971).

30. F. Crick, cited in *Nature (Lond.)* 220, 429 (1968).

31. A. Etzioni, *Science* 161, 1107 (1968).

32. A. G. Motulsky, *Israel J. Med. Sci.* 9, 1341 (1973).

33. G. Stamatoyannopoulos, in *Proceedings of the Fourth International Conference on Birth Defects,* Vienna, Austria, September 1973, in press.

34. M. M. Kaback and J. S. O'Brien, in *Medical Genetics,* V. McKusick and R. Claiborne, Eds. (HP Publishing, New York, 1973).

35. Testing for Tay-Sachs carrier status requires an assay of hexosaminidase A. The level of this enzyme in the plasma increases during normal pregnancy. However, white cells can still be used to discriminate between normal and carrier pregnant women. Carrier testing during pregnancy, therefore, would be technically more difficult but probably could be worked out logistically.

36. J. L. Goldstein, H. G. Schrott, W. R. Hazzard, E. L. Bierman, A. G. Motulsky, *J. Clin, Invest.* 52, 1544 (1973).

37. E. B. Hook, *Science* 179, 139 (1973).

38. H. Muller, *Perspect. Biol. Med.* 3, 1 (1959).

39. L. R. Kass, *N. Engl. J. Med.* 285, 1174 (1971).

40. R. G. Edwards, in *The Biological Revolution: Social Good or Social Evil?* W. Fuller, Ed. (Doubleday, New York, 1972), chap, 9, p. 128; in *Proceedings of the Fourth International Conference on Birth Defects,* Vienna, Austria, September 1973, in press.

41. Editorial, *J. Am. Med. Assoc.* 220 (No 5), 721 (1972).

42. J. Lederberg, in *Challenging Biological Problems,* J. A. Behnke, Ed. (Oxford Univ. Press, New York, 1972), chap. 1, p. 7; *Am. Nat.* **100**, 519 (1966).

43. B. D. Davis, *Science* **170**, 1279 (1970); P. Ramsey, *Fabricated Man* (Yale Univ. Press, New Haven, Conn., 1970).

44. T. Friedmann and R. Roblin, *Science* **175** 949 (1972).

45. E. Freese, Ed., *The Prospects of Gene Therapy* [Fogarty International Center Conference Report, Department of Health, Education, and Welfare, Publ. No. (NIH) 72-61, 1972].

46. Supported in part by PHS grant GM-15253.

Bibliography

The references cited here will provide the interested reader with sources for further reading about basic genetics, human genetics, and the bioethical problems associated with applying advances in genetics to humans. In addition to these sources, many excellent references are cited in the articles included in this book.

Augenstein, L. G. 1968. *Come, Let Us Play God.* Harper and Row, New York.

Bergsma, D. (ed) 1971. *Symposium on Intrauterine Diagnosis.* Birth Defects: Original Article Series, Vol VII, No. 5. The National Foundation-March of Dimes. White Plains, New York.

Bergsma, D. (ed) 1972. *Advances in Human Genetics and Their Impact on Society.* Birth Defects: Original Articles Series, Vol. VIII, No. 4. The National Foundation–March of Dimes. White Plains, New York.

Bergsma, D. (ed) 1973. *Contemporary Genetic Counseling.* Birth Defects: Original Articles Series, Vol. IX, No. 4. The National Foundation–March of Dimes. White Plains, New York.

Bergsma, D. (ed) 1974. *Ethical, Social and Legal Dimensions of Screening for Human Genetic Disease.* Birth Defects: Original Article Series. Vol. X, No. 6. The National Foundation–March of Dimes. White Plains, New York.

Fletcher, Joseph. 1974. *The Ethics of Genetic Control. Ending Reproductive Roulette.* Anchor/Doubleday, New York.

Fuhrmann, Walter, and Friedrich Vogel, 1969. *Genetic Counseling.* Springer-Verlag, New York.

Gardner, E. J. 1975. *Principles of Genetics,* Fifth Edition. John Wiley, New York.

Grobman, A. B. (ed.) 1970. *Social Implications of Biological Education.* National Association of Biology Teachers, Washington, D. C.

Hamilton, Michael (ed.) 1972. *The New Genetics and the Future of Man.* Wm. B. Eerdmans, Grand Rapids, Michigan.

Lederberg, J. 1972. "Biological Innovation and Genetic Intervention" in *Challenging Biological Problems. Directions Toward Their Solution.* Edited by J. A. Behnke. Oxford University Press, New York.

Lerner, I. M. 1968. *Heredity, Evolution and Society.* W. H. Freeman, San Francisco.

Levitan, Max and Ashley Montagu. 1971. *Textbook of Human Genetics.* Oxford University Press, New York.

McKusick, V. A. 1969. *Human Genetics,* Second Edition. Prentice-Hall, Englewood Cliffs, New Jersey.

Nagle, J. J. 1974. *Heredity and Human Affairs.* C. V. Mosby, St. Louis.

Pai, A. C. 1974. *Foundations of Genetics. A Science for Society.* McGraw-Hill, New York.

Porter, I. H. and R. G. Skalko (eds.) 1973. *Heredity and Society.* Academic Press, New York.

Potter, V. R. 1971. *Bioethics. Bridge to the Future.* Prentice-Hall, Englewood Cliffs, New Jersey.

Ramsey, Paul. 1970. *Fabricated Man—The Ethics of Genetic Control.* Yale University Press, New Haven, Connecticut.

Roslansky, J. D. (ed.) 1966. *Genetics and the Future of Man.* Appleton-Century-Crofts, New York.

Sonneborn, T. M. (ed.) 1965. *The Control of Human Heredity and Evolution.* Macmillan, New York.

Stern, C. 1973. *Principles of Human Genetics.* W. H. Freeman, San Francisco.

Volpe, E. P. 1971. *Human Heredity and Birth Defects.* Pegasus, New York.

Winchester, A. M. 1971. *Human Genetics.* Charles E. Merrill, Columbus, Ohio.

Young, L. B. (ed.) 1970. *Evolution of Man.* Oxford University Press, New York.